Everyday Mathematics®

The University of Chicago School Mathematics Project

STUDENT MATH JOURNAL

VOLUME 1

Bothell, WA • Chicago, IL • Columbus, OH • New York, NY

The University of Chicago School Mathematics Project

Max Bell, Director, *Everyday Mathematics* First Edition; James McBride, Director, *Everyday Mathematics* Second Edition; Andy Isaacs, Director, *Everyday Mathematics* Third, CCSS, and Fourth Editions; Amy Dillard, Associate Director, *Everyday Mathematics* Third Edition; Rachel Malpass McCall, Associate Director, *Everyday Mathematics* CCSS and Fourth Editions; Mary Ellen Dairyko, Associate Director, *Everyday Mathematics* Fourth Edition

Authors
Max Bell, Jean Bell, John Bretzlauf, Amy Dillard, Robert Hartfield, Andy Isaacs, James McBride, Cheryl G. Moran, Kathleen Pitvorec, Peter Saecker

Fourth Edition Grade 2 Team Leader
Cheryl G. Moran

Writers
Camille Bourisaw, Mary Ellen Dairyko, Gina Garza-Kling, Rebecca Williams Maxcy, Kathryn M. Rich

Open Response Team
Catherine R. Kelso, Leader; Steve Hinds

Differentiation Team
Ava Belisle-Chatterjee, Leader; Jean Marie Capper

Digital Development Team
Carla Agard-Strickland, Leader; John Benson, Gregory Berns-Leone, Juan Camilo Acevedo

Virtual Learning Community
Meg Schleppenbach Bates, Cheryl G. Moran, Margaret Sharkey

Technical Art
Diana Barrie, Senior Artist; Cherry Inthalangsy

UCSMP Editorial
Don Reneau, Senior Editor; Rachel Jacobs, Kristen Pasmore, Luke Whalen

Field Test Coordination
Denise A. Porter

Field Test Teachers
Kristin Collins, Debbie Crowley, Brooke Fordice, Callie Huggins, Luke Larmee, Jaclyn McNamee, Vibha Sanghvi, Brook Triplett

Contributors
William B. Baker, John Benson, James Flanders, Lila K. S. Goldstein, Funda Gönülateş, Lorraine M. Males, John P. Smith III, Kathleen Clark, Patti Satz, Penny Williams

Center for Elementary Mathematics and Science Education Administration
Martin Gartzman, Executive Director; Jose J. Fragoso, Jr., Meri B. Forhan, Regina Littleton, Laurie K. Thrasher

External Reviewers
The *Everyday Mathematics* authors gratefully acknowledge the work of the many scholars and teachers who reviewed plans for this edition. All decisions regarding the content and pedagogy of *Everyday Mathematics* were made by the authors and do not necessarily reflect the views of those listed below.

Elizabeth Babcock, California Academy of Sciences; Arthur J. Baroody, University of Illinois at Urbana-Champaign and University of Denver; Dawn Berk, University of Delaware; Diane J. Briars, Pittsburgh, Pennsylvania; Kathryn B. Chval, University of Missouri-Columbia; Kathleen Cramer, University of Minnesota; Ethan Danahy, Tufts University; Tom de Boor, Grunwald Associates; Louis V. DiBello, University of Illinois at Chicago; Corey Drake, Michigan State University; David Foster, Silicon Valley Mathematics Initiative; Funda Gönülateş, Michigan State University; M. Kathleen Heid, Pennsylvania State University; Natalie Jakucyn, Glenbrook South High School, Glenview, IL; Richard G. Kron, University of Chicago; Richard Lehrer, Vanderbilt University; Susan C. Levine, University of Chicago; Lorraine M. Males, University of Nebraska-Lincoln; Dr. George Mehler, Temple University and Central Bucks School District, Pennsylvania; Kenny Huy Nguyen, North Carolina State University; Mark Oreglia, University of Chicago; Sandra Overcash, Virginia Beach City Public Schools, Virginia; Raedy M. Ping, University of Chicago; Kevin L. Polk, Aveniros LLC; Sarah R. Powell, University of Texas at Austin; Janine T. Remillard, University of Pennsylvania; John P. Smith III, Michigan State University; Mary Kay Stein, University of Pittsburgh; Dale Truding, Arlington Heights District 25, Arlington Heights, Illinois; Judith S. Zawojewski, Illinois Institute of Technology

Note
Too many people have contributed to earlier editions of *Everyday Mathematics* to be listed here. Title and copyright pages for earlier editions can be found at http://everydaymath.uchicago.edu/about/ucsmp-cemse/.

www.everydaymath.com

Send all inquiries to:
McGraw-Hill Education
STEM Learning Solutions Center
8787 Orion Place
Columbus, OH 43240

ISBN: 978-0-02-143082-6
MHID: 0-02-143082-9

Printed in the United States of America.

3 4 5 6 7 8 9 RMN 19 18 17 16 15 14

Contents

Unit 1

Unit 2

Unit 3

Unit 4

Activity Sheets

Counts on a Number Line

Lesson 1-1

DATE

Fill in the missing numbers.

1

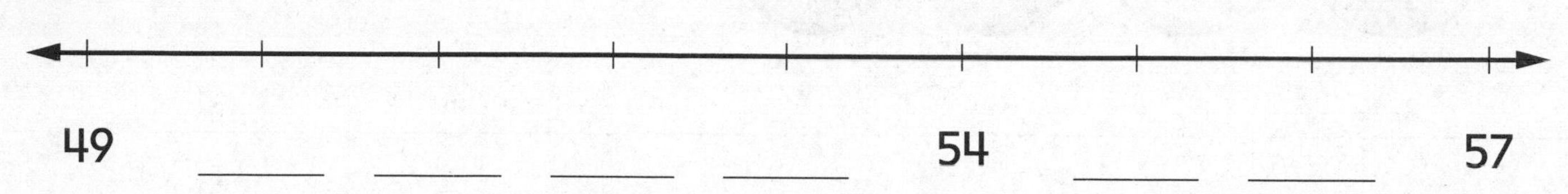

2

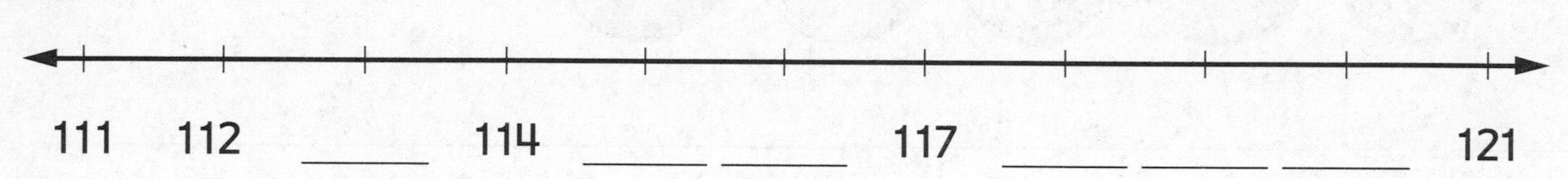

3

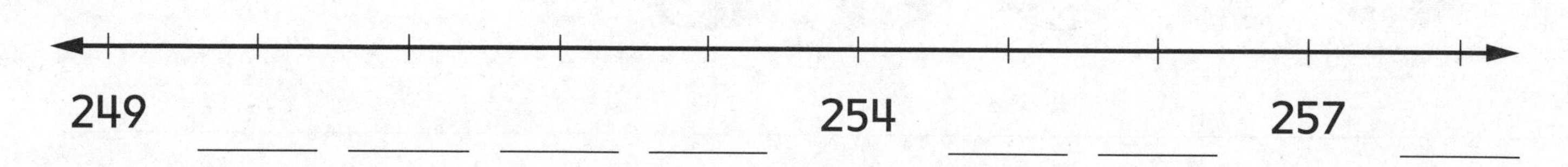

4

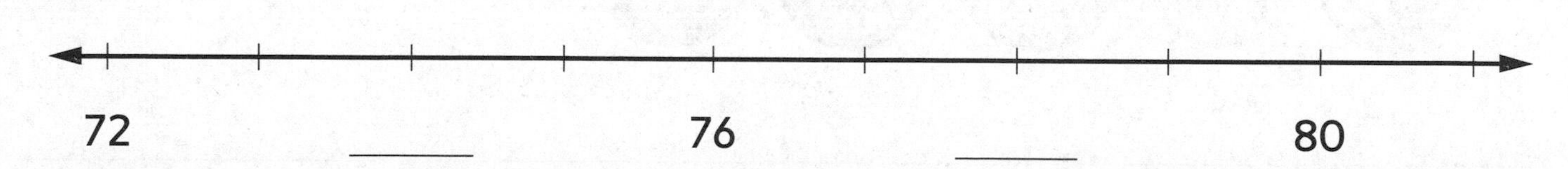

Try This

5 Place the number 5 in the correct spot on the number line.

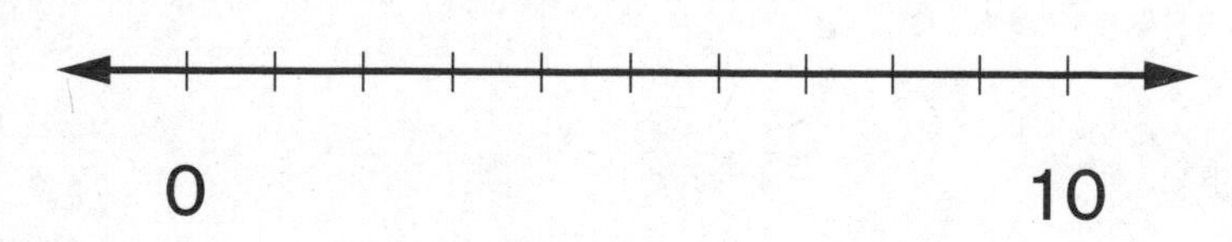

Coins

Lesson 1-3

DATE

= ______¢

= ______¢

= ______¢

= ______¢

= ______¢

Coin Pattern

Lesson 1-5

DATE

Look for a pattern in the coins.

1. What will be the next coin in the pattern?

2. Explain to a partner the pattern you noticed that helped you figure out the next coin.

Broken Calculator

Lesson 1-6

DATE

Example: Show 17. Broken key is (7). Show several ways: 11 + 6 20 − 3 8 + 8 + 1	**1** Show 3. Broken key is (3). Show several ways: __________ __________ __________
2 Show 20. Broken key is (2). Show several ways: __________ __________ __________	**3** Show 22. Broken key is (2). Show several ways: __________ __________ __________
4 Show 12. Broken key is (1). Show several ways: __________ __________ __________	**5** Make up your own. Show ________. Broken key is ________. Show several ways: __________ __________ __________

Missing Numbers on Number Lines

Lesson 1-7

DATE

1. Place the numbers 1, 3, and 8 in the correct spots on the number line.

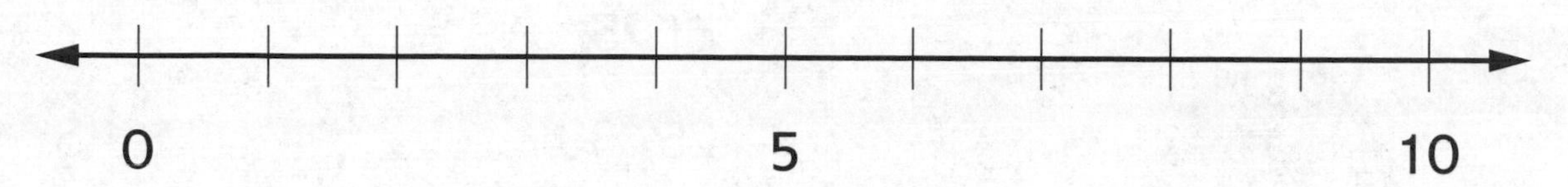

2. Place the numbers 23 and 26 in the correct spots on the number line.

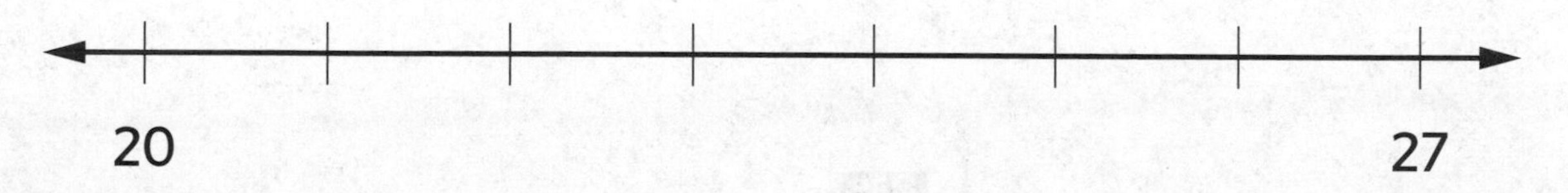

3. Place the numbers 86 and 88 in the correct spots on the number line.

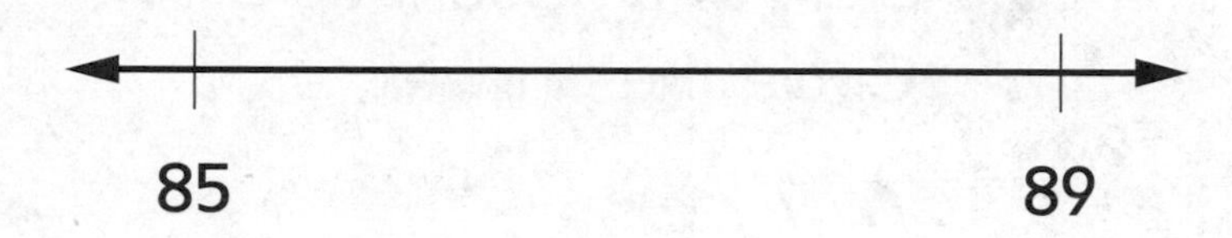

4. Look at Number Line A and Number Line B. What do you notice about the distance between 32 and 33 on each number line?

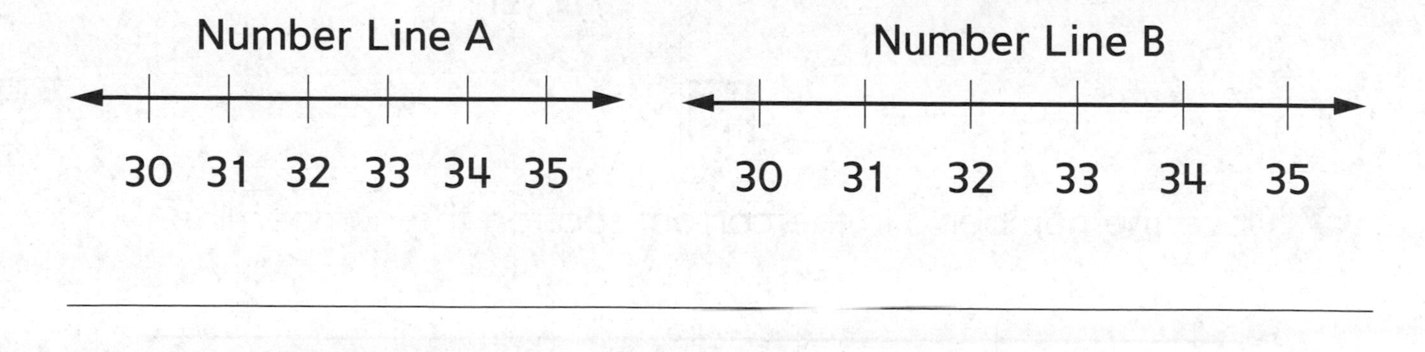

Math Boxes

Lesson 1-8

DATE

Math Boxes

1. How many cents?

_____¢

2. Skip count by 5s beginning at 25.

25, 30, _____, _____, 45,

_____, _____, 60

MRB 67

3. Show two ways to make 25 cents.

Use Ⓟ, Ⓝ, and Ⓓ.

4. Start at 12. Count up 5. Circle the answer.

A. 7

B. 17

C. 18

D. 21

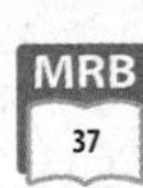

5. Place the number 3 in the correct spot on the number line.

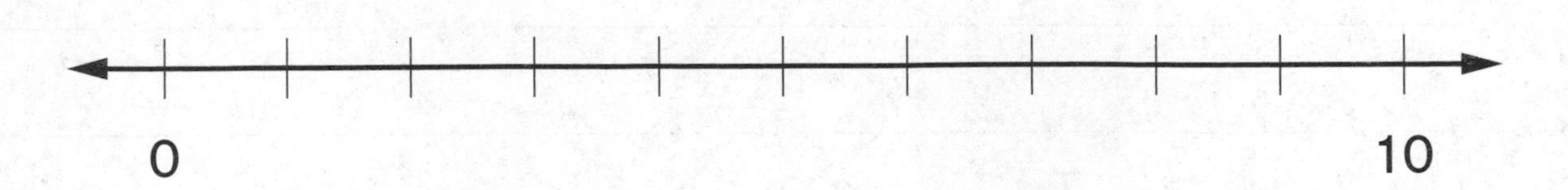

Odd and Even Patterns

Lesson 1-9

DATE

Write the number of dots. Label the number odd or even.

Example:

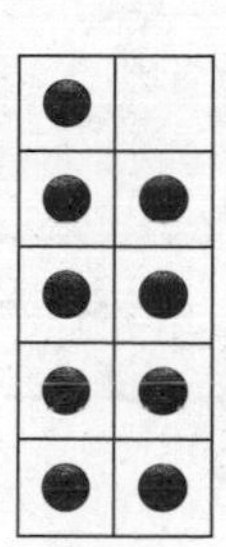

9

Odd

1

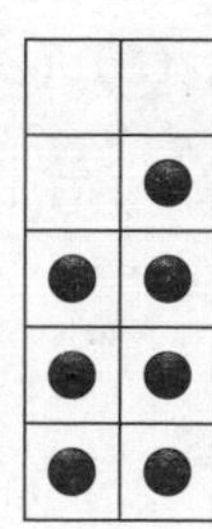

______

____

2

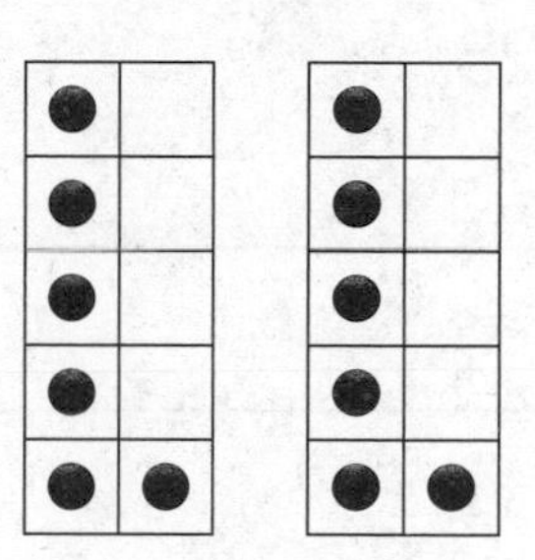

______

____

3

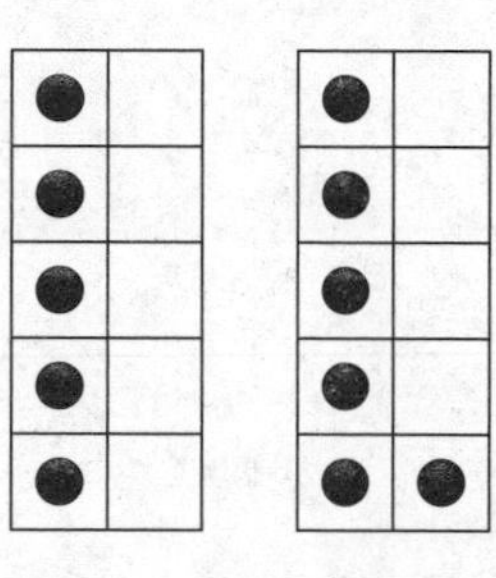

______

____

4 Write an even number. ______
Show how you know the number is even.

5 Write an odd number. ______
Show how you know the number is odd.

Math Boxes

1 How much money is shown?

Ⓓ Ⓓ Ⓓ Ⓓ

_____ ¢

MRB 110-111

2 Fill in the missing numbers.

	4			7
13		15		

MRB 67

3 Count the tally marks.

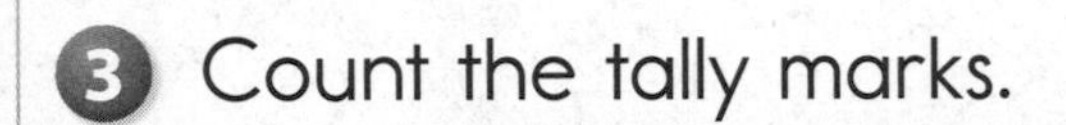

There are _____ tally marks.

4 Use your calculator.

Pretend the ⑤ key is broken.

Write four different ways to show 25.

____

____

____

____

5 **Writing/Reasoning** Explain how you figured out the missing numbers in Problem 2.

____

____

____

____

Finding Patterns on a Number Grid

Lesson 1-10

DATE

1. Start at 102 and count by 5s. Color the counts.
   You may use your calculator.

									100
101	102	103	104	105	106	107	108	109	110
111	112	113	114	115	116	117	118	119	120
121	122	123	124	125	126	127	128	129	130
131	132	133	134	135	136	137	138	139	140
141	142	143	144	145	146	147	148	149	150

2. Write the shaded numbers in order on the lines.
   Circle the digits in the ones place.

   ______, ______, ______, ______, ______,

   ______, ______, ______, ______, ______

3. What pattern do you notice in the numbers you wrote for Problem 2?

   ____

   ____

Try This

4. What digits would be in the ones place if you counted by 5s and started at 103? ______________ Explain how you know.

   ____

   ____

   ____

DATE

Broken Calculator

1. Show 8.
Broken key is 8.
Show several ways:

____

____

____

2. Show 15.
Broken key is 5.
Show several ways:

____

____

____

3. Show 26.
Broken key is 6.
Show several ways:

____

____

____

4. Show 30.
Broken key is 3.
Show several ways:

____

____

____

5. Make up your own.
Show __________.
Broken key is __________.
Show several ways:

____

____

____

6. Make up your own.
Show __________.
Broken key is __________.
Show several ways:

____

____

____

1 Write 6 names in the 10 box.

10

MRB 53

2 Solve.

Unit: Birds

5 + 5 = 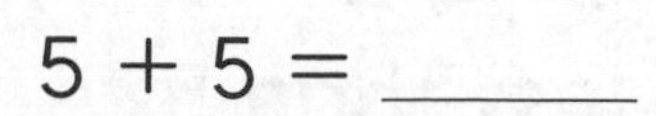

______ = 4 + 6

3 + ______ = 10

______ + 9 = 10

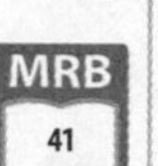

3 Write +, −, or =.

Unit: Pencils

10 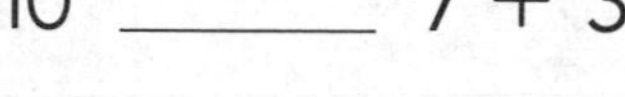 7 + 3

4 ______ 7 − 3

7 ______ 3 = 10

7 ______ 3 = 4

MRB 36

4 Write three doubles facts.

____

____

____

MRB 40

5 Use the number line to solve 5 + 7. Show your work.

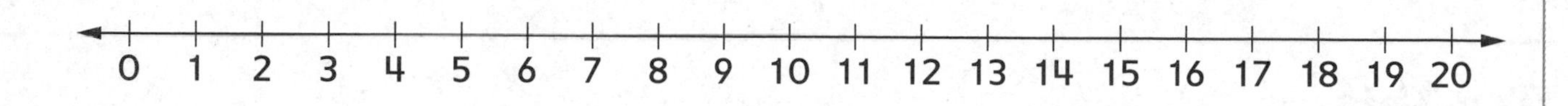

5 + 7 = ______

MRB 37

Using <, >, and =

Lesson 1-11

DATE

3 < 5 — 3 is less than 5.

5 > 3 — 5 is greater than 3.

Write <, >, or =.

1. 18¢ ______ 81¢
2. 61 ______ 16
3. 107 ______ 107
4. $1 ______ 94¢
5. 299 ______ 301
6. 1,032 ______ 1,132

Write the total amounts. Then write <, >, or =.

Example: Ⓓ Ⓝ Ⓝ Ⓟ Ⓟ = 22¢ < 26¢ = Ⓠ Ⓟ

7. Ⓝ Ⓝ Ⓓ Ⓟ = ______¢ ______ ______¢ = Ⓠ Ⓝ Ⓓ Ⓝ

8. Ⓓ Ⓓ Ⓠ Ⓓ = ______¢ ______ ______¢ = Ⓓ Ⓝ Ⓟ Ⓓ

Try This

9. 15 ______ 7 + 8
10. 5 + 6 ______ 8 + 4

Math Boxes

Lesson 1-11

DATE

1. How much money is shown?

Ⓓ Ⓓ Ⓓ Ⓓ Ⓓ Ⓓ

______¢

MRB 110-111

2. Fill in the missing numbers.

2			
	13		15

MRB 67

3. Show 23 with tally marks.

MRB 113

4. Use your calculator to show 14.
Broken key is ①.
Show four different ways.

______

______

______

______

5. **Writing/Reasoning** How did you count the money in Problem 1?

______

______

______

______

MRB 18-19, 67

Covering Rectangles with Shapes

Lesson 1-12

DATE

Cover this rectangle with squares.

Cover this rectangle with triangles.

Cover this rectangle with circles.

Math Boxes

Lesson 1-12

DATE

1. How much money is shown?

Ⓓ Ⓓ Ⓟ Ⓟ Ⓟ Ⓟ Ⓟ

_____¢

2. Skip count by 5s beginning at 20.

20, _____, _____, 35,

_____, _____, 50

3. Show two ways to make 30 cents.

Use Ⓟ, Ⓝ, and Ⓓ.

MRB 110-111

4. Start at 15. Count up 9. Circle the answer.

A. 22

B. 24

C. 25

D. 34

5. Write the number 6 in the correct spot on number line.

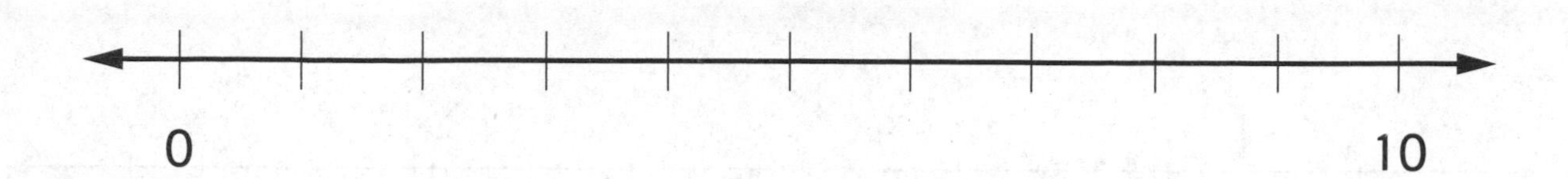

Math Boxes
Preview for Unit 2

Lesson 1-13

DATE

1. Write 6 addition names in the 10 box.

   10

   MRB 53

2. Solve.

   Unit: birds

   $8 + 2 =$ ______

   ______ $= 6 + 4$

   $7 +$ ______ $= 10$

   ______ $+ 1 = 10$

   MRB 41

3. Write +, −, or =.

   Unit: birds

   5 ______ $5 = 10$

   5 ______ $5 = 0$

   8 ______ $2 = 6$

   8 ______ $2 = 10$

   MRB 36

4. Write at least three double facts you know.

   ______

   ______

   ______

   MRB 40

5. Use the number line to solve 7 + 8. Show your work.

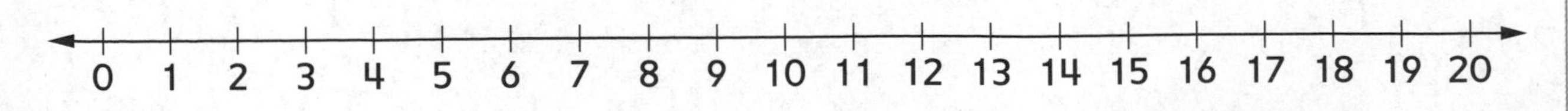

   $7 + 8 =$ ______

   MRB 37

Counting Money

Lesson 2-1

DATE

Write the amount.

1.   

= $______

2.   
= $______

3.    

= $______

4.    

= $______

Math Boxes

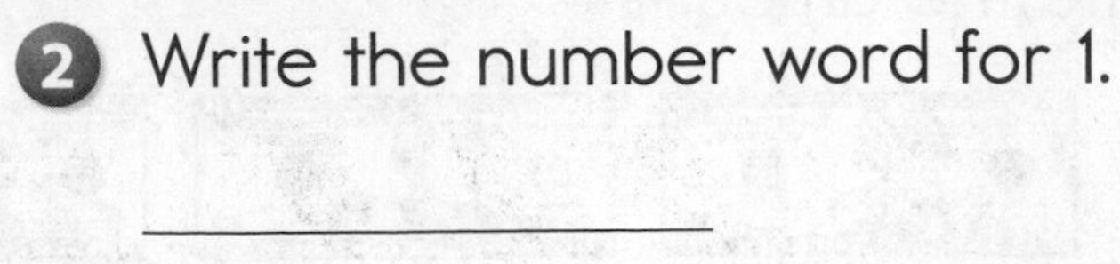

Math Boxes

1. Count by 10s.

   ______, 10, ______, ______,

   40, ______, ______

2. Write the number word for 1.

   ____

3. Is the total number of dots odd or even?

   ____

4. Write the missing numbers.

   97, 98, ______, ______,

   ______, 102

5. **Writing/Reasoning** Look back at Problem 1. What patterns do you notice in the numbers?

   ____

   ____

   ____

Number Stories

Lesson 2-2

DATE

Write an addition number story about what you see in the picture.
Write a label in the unit box. Find the answer. Write a number model.

Example: *There are 7 ducks in the water. There are 3 ducks in the grass. How many ducks are there in all?*

Answer: *10 ducks*

Number model: *7* + *3* = *10*

Unit
ducks

Story: ______________________________

____

____

____

Answer: ______________

Number model: ______ + ______ = ______

Unit
children

Number-Grid Puzzles

Lesson 2-2

DATE

Complete the number-grid puzzles.

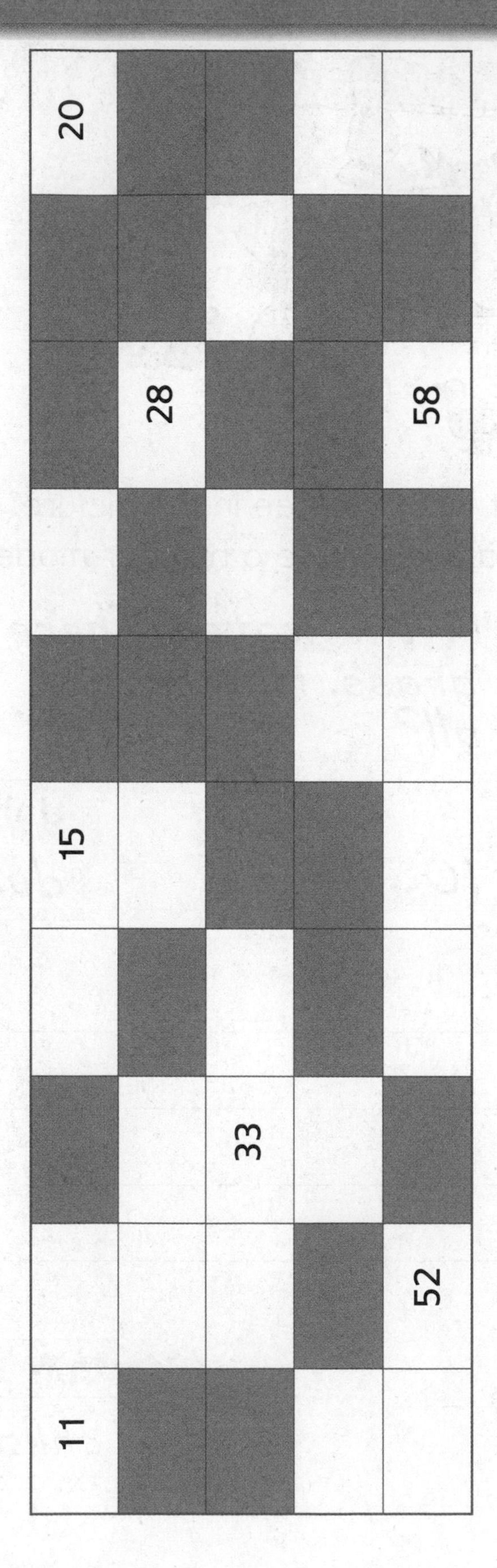

11				15					20
							28		
		33							
	52						58		

	332				336				
								349	
						357			
361									
		373					378		

Math Boxes

DATE

1 Use a number grid. How many spaces from

18 to 28? ______

49 to 59? ______

2 Show \$1.
Use Ⓟ, Ⓝ, Ⓓ, and Ⓠ.

3 Use <, > or =.

10 ______ 9 + 3

12 ______ 8 + 2

6 + 3 ______ 9

7 + 3 ______ 11

4 Which numbers are larger than 38? Fill in the circles next to correct answers.

Ⓐ 36
Ⓑ 87
Ⓒ 70
Ⓓ 59

5 Count by 25s.

0, 25, ______, ______,

100, ______, 150

6 There are 6 pencils in a cup. Then 3 more are added. How many pencils are now in the cup?

Unit
pencils

Answer: ______ pencils

Number model:

____

MRB 27

My Addition Fact Strategies

DATE

Use this page to record addition strategies. When you don't know how to solve an addition problem, you can look here for ideas.

Strategy:	Strategy:
Strategy:	Strategy:
Strategy:	Strategy:
Strategy:	Strategy:

Math Boxes

1. Count by 5s.

   55, ______, ______, 70, ______,

   ______, 85

   MRB 67

2. Write the number word for 3.

   ____

3. Is the total number of dots odd or even?

   ● ●
   ● ●
   ● ●
   ● ●
   ● ●

   ____

   MRB 59-60

4. Write the missing numbers.

   ______, ______, 119, ______,

   ______, 122

5. **Writing/Reasoning** In Problem 1, what patterns do you notice in the numbers?

   ____

   ____

   ____

Practicing the Making-10 Strategy

Lesson 2-4

DATE

Use the double ten frames to make 10. Then find the sum.
Write the combination of 10 that helped.

Example:

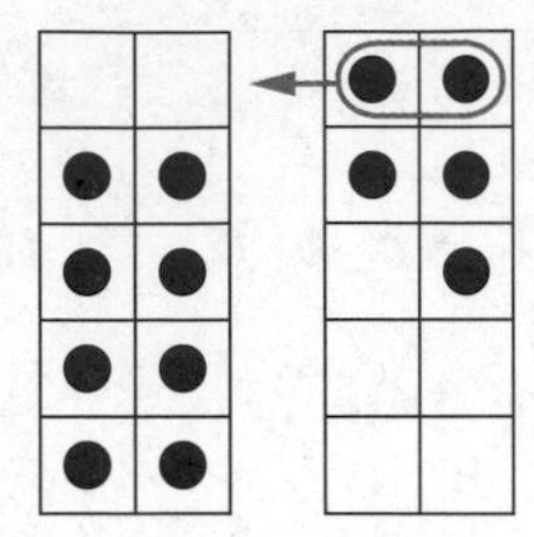

Combination of 10 that helped:

8 + 2 = 10

Fact: 8 + 5 = 13

1

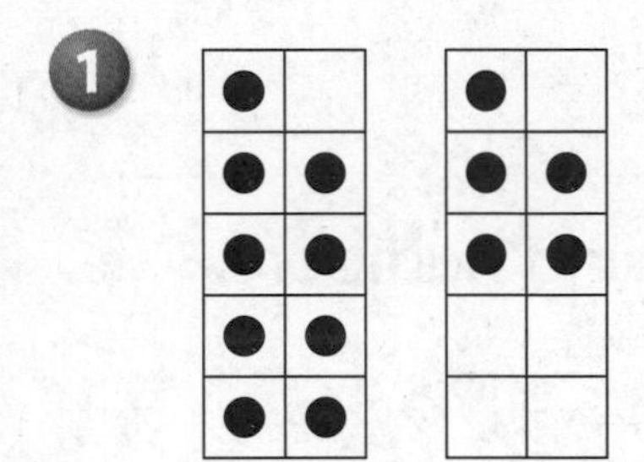

Combination of 10 that helped:

_____ + _____ = _____

Fact: _____ + _____ = _____

2

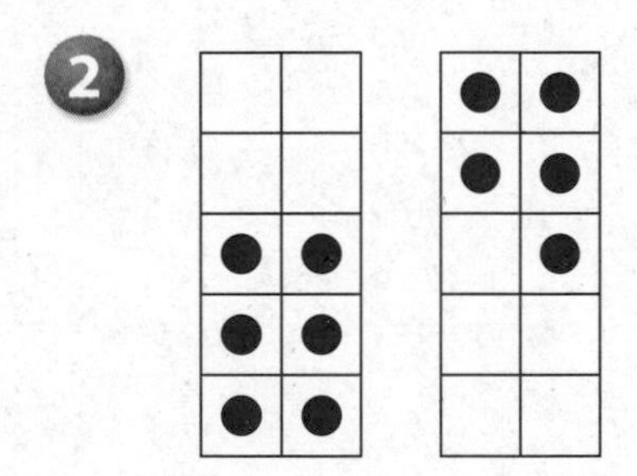

Combination of 10 that helped:

_____ + _____ = _____

Fact: _____ + _____ = _____

3 Explain how you solved Problem 2.

Math Boxes

Lesson 2-4

DATE

1. Use a number grid. How many spaces from

17 to 26? ______

49 to 65? ______

2. Show 88¢.
Use Ⓟ, Ⓝ, Ⓓ, and Ⓠ.

3. Use <, >, or =.

8 + 3 ______ 11

10 ______ 9 + 2

11 ______ 6 + 3

9 + 3 ______ 8

4. Which numbers are smaller than 38? Fill in the circles next to the correct answers.

Ⓐ 41

Ⓑ 19

Ⓒ 33

Ⓓ 39

5. Count by 25s.

0, 25, ______, 75, ______,

125, ______

6. Eight children are playing a game. Three more come to play. How many children are playing in all?

Unit
children

Answer: ______ children

Number model:

____

MRB 27

Math Boxes

Helper Doubles Facts

Lesson 2-5

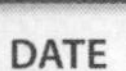

DATE

For each problem, do the following:

- Circle the helper doubles in the ten frames.
- Write the helper doubles fact.
- Write the fact that matches the double ten frames.

Example:

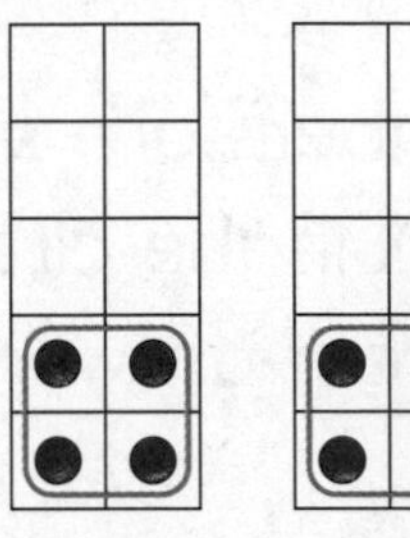

Helper doubles fact:

4 + 4 = 8

Fact: 4 + 5 = 9

1

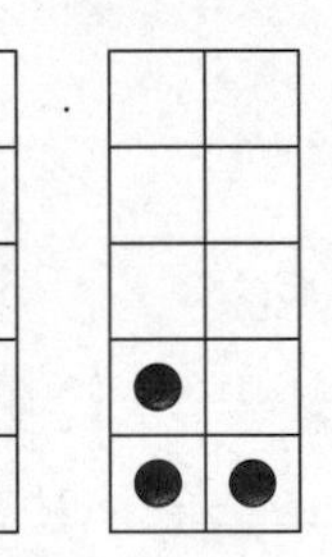

Helper doubles fact:

_____ + _____ = _____

Fact: _____ + _____ = _____

2

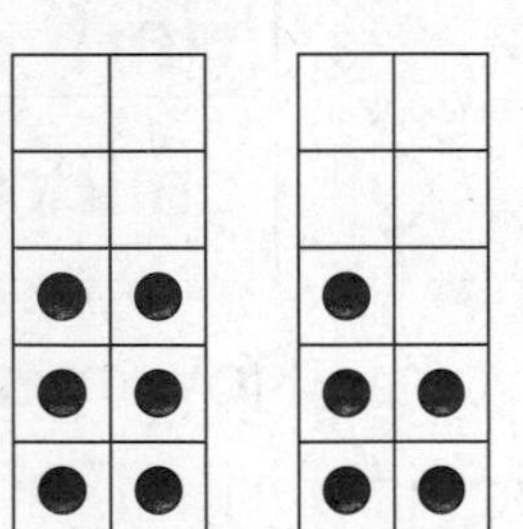

Helper doubles fact:

_____ + _____ = _____

Fact: _____ + _____ = _____

3

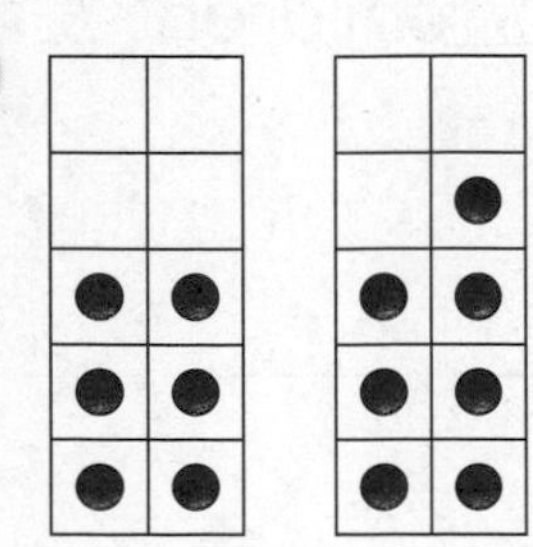

Helper doubles fact:

_____ + _____ = _____

Fact: _____ + _____ = _____

DATE

Help Doubles Facts (continued)

4 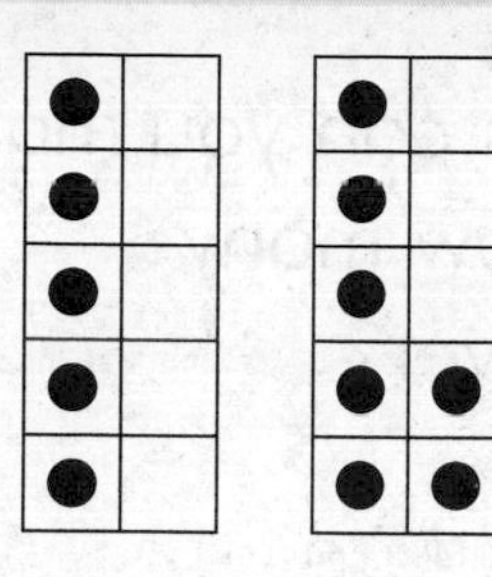

Helper doubles fact:

_____ + _____ = _____

Fact: _____ + _____ = _____

Try This

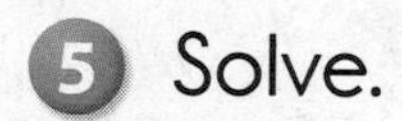

5 Solve.

8 + 6 = _____

How did you figure out the answer?

____

____

____

____

____

____

DATE

Math Boxes

1. How much money is shown? Fill in the correct answer.

   Ⓓ Ⓓ Ⓓ Ⓝ Ⓟ

   ◯ 41¢

   ◯ 36¢

   ◯ 32¢

   ◯ 5¢

   MRB 110-111

2. How many tens can you make? Circle them. How many ones are left over?

   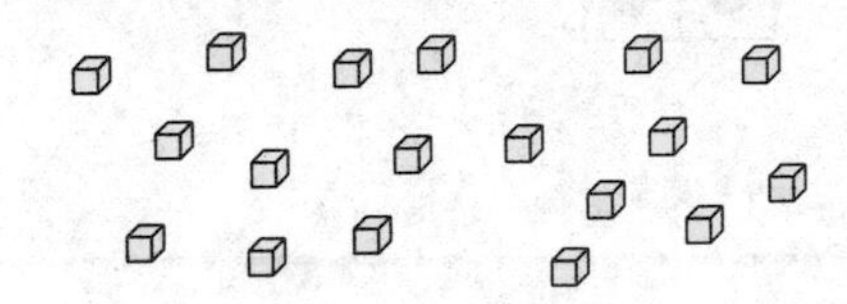

   _____ ten

   _____ ones

   MRB 70

3. Use your calculator to show 18.

   The [8] key is broken.

   Show four ways:

   _____

   _____

   _____

   _____

4. There were 4 frogs in the pond. More frogs jump into the pond. Now there are 10 frogs. How many frogs just jumped into the pond?

Unit
frogs

   Answer: _____ frogs

   Number model:

   _____

   MRB 27

5. Place the number 111 in the correct spot on the number line.

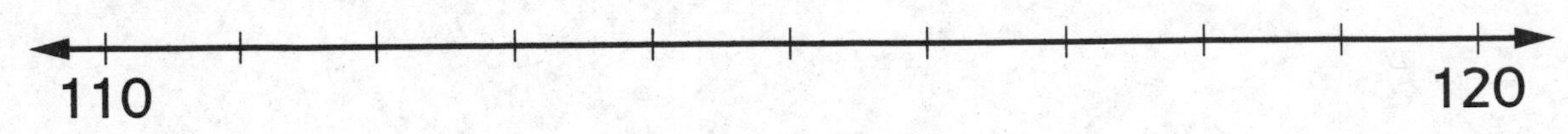

Even or Odd?

Lesson 2-6

DATE

Work with a partner. For Problems 1–3, do the following:

- Write the number of pennies you grabbed and the number of pennies your partner grabbed.
- Write the number of pennies you have in all.
- Circle "even" or "odd" to show whether each number is even or odd.

1. Your grab: ______ even odd

   Your partner's grab: ______ even odd

   Total: ______ even odd

2. Your grab: ______ even odd

   Your partner's grab: ______ even odd

   Total: ______ even odd

3. Your grab: ______ even odd

   Your partner's grab: ______ even odd

   Total: ______ even odd

4. Draw a picture or use words to explain how you know when a number is even.

____

____

DATE

Math Boxes

Math Boxes

1. Fill in the missing numbers.

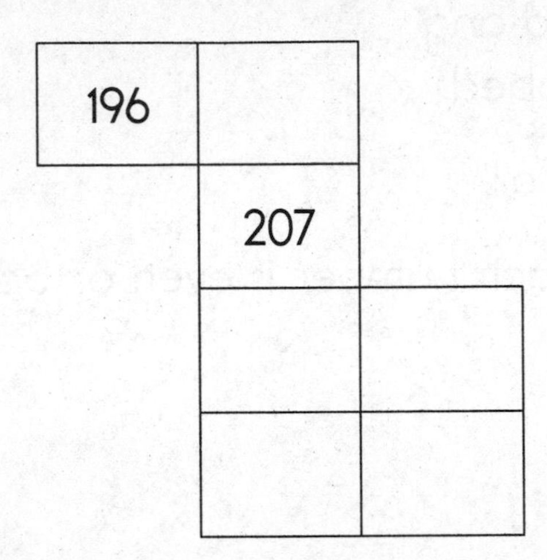

2. Write <, >, or =.

92 _____ 89

115 _____ 117

26 _____ 60

105 _____ 103

< is less than
> is more than
= is the same as

MRB 74-75

3. Fill in the missing numbers.

_____, 129, _____, _____,

132, _____, _____

4. Show two ways to make 30¢.
Use Ⓟ, Ⓝ, and Ⓓ.

5. **Writing/Reasoning** Pick one answer from Problem 2.
Explain how you know which number is greater.

____

____

____

Number Stories

Lesson 2-7

DATE

1. Write a number model for this number story.

   Jessica has 2 dogs and 8 goldfish. How many pets does she have in all?

   ____

   Talk with a partner about what makes a story a number story.

Math Boxes

1. How much money is shown?

   Ⓠ Ⓓ Ⓓ Ⓟ Ⓟ Ⓟ

   _______ ¢

   MRB 110-111

2. How many tens can you make? Circle them. How many ones are left over?

   _______ tens

   _______ ones

   MRB 70

3. Use your calculator to show 15.

   The [5] key is broken.

   Show four ways:

   ____

   ____

   ____

   ____

4. There were 6 birds in a nest. More birds flew to the nest. Then there were 10 birds. How many birds flew to the nest?

Unit
birds

   - A. 16 birds
   - B. 8 birds
   - C. 5 birds
   - D. 4 birds

5. Place the number 138 in the correct spot on the number line.

   130 ———————————————— 140

Using Tools to Add

Lesson 2-8

DATE

Use the number lines below to find the sums. Then find the sums using the Number Grid page.

1. Show 62 + 10 on the number line below.

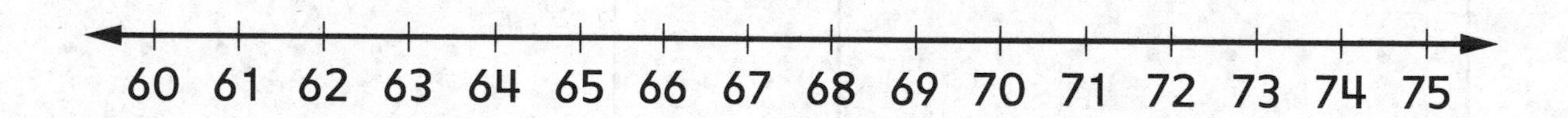

Answer: ______

Show 62 + 10 on the number grid. Draw arrows to show your counts.

2. Show 39 + 10 on the number line below.

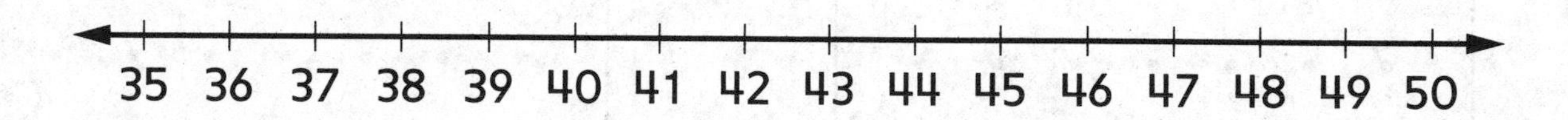

Answer: ______

Show 39 + 10 on the number grid. Draw arrows to show your counts.

Try This

3. Show 45 + 12 on the number line below.

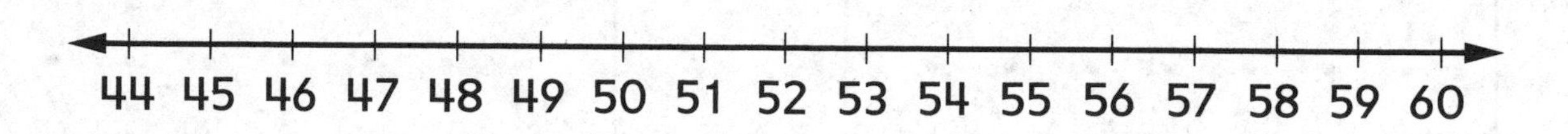

Answer: ______

Show 45 + 12 on the number grid. Draw arrows to show your counts.

DATE

Geoboard Dot Paper (7 by 7)

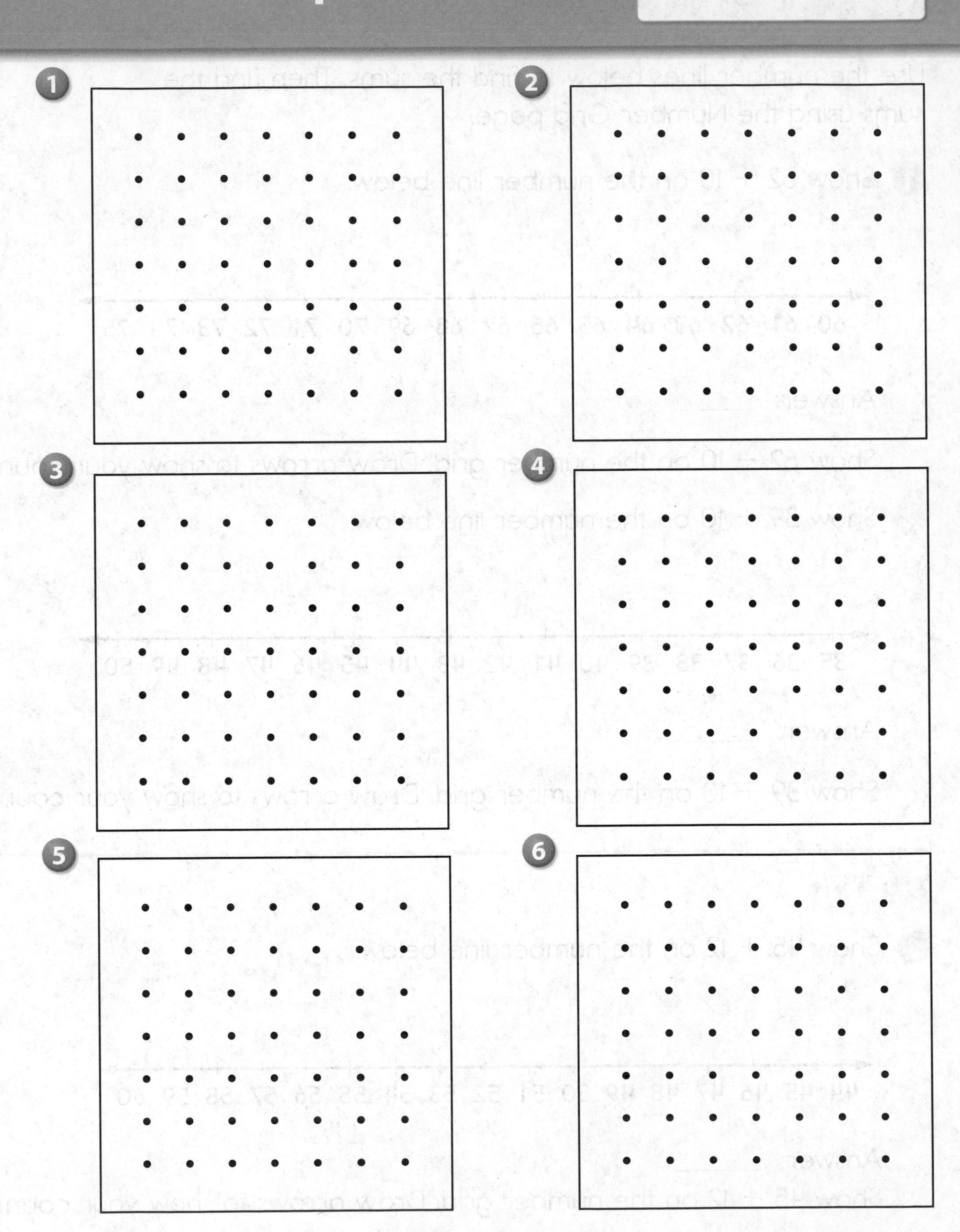

Math Boxes

1 Fill in the missing numbers.

144		
	155	

2 Write <, >, or =.

26 ______ 36

87 ______ 78

111 ______ 101

354 ______ 345

< is less than
> is more than
= is the same as

MRB 74-75

3 Fill in the missing numbers.

148, ______, ______, ______,

152, 153, ______

4 Show two ways to make 35¢.
Use Ⓟ, Ⓝ, Ⓓ, and Ⓠ.

MRB 110-111

5 **Writing/Reasoning** Show 35¢ using as few coins as possible.

Explain how you know you used the smallest number of coins.

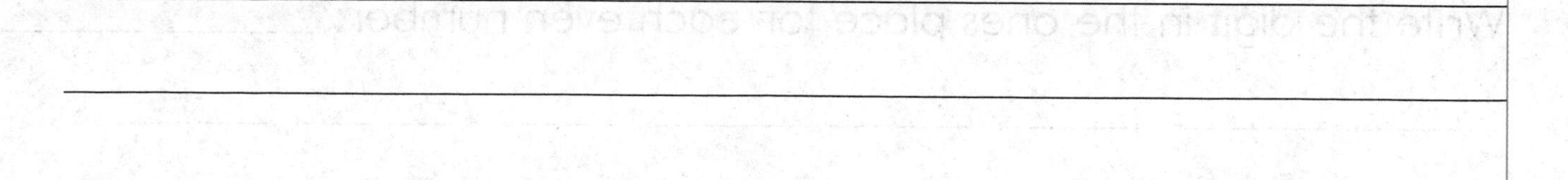

Exploring Even and Odd Numbers

Lesson 2-9

DATE

1. With a partner to sort your number cards. Make a pile of even numbers and a pile of odd numbers. Use counters to help.
2. Put the even numbers in order from smallest to largest.
3. Put the odd numbers in order from smallest to largest.
4. Record your numbers below.
5. For each number, circle the digit in the ones place.

Odd Numbers	Even Numbers

Write the digit in the ones place for each even number: ______, ______,

______, ______, ______, ______, ______, ______, ______, ______

Write the digit in the ones place for each odd number: ______, ______,

______, ______, ______, ______, ______, ______, ______, ______

1 Write the fact family for this Fact Triangle.

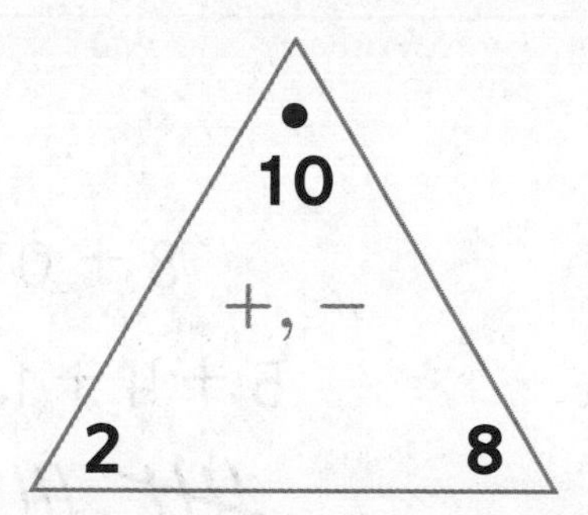

$8 + 2 =$ ______ $10 - 2 = 8$

$2 + 8 =$ ______ $10 - 8 = 2$

MRB 47

2 Write the fact family for this Fact Triangle.

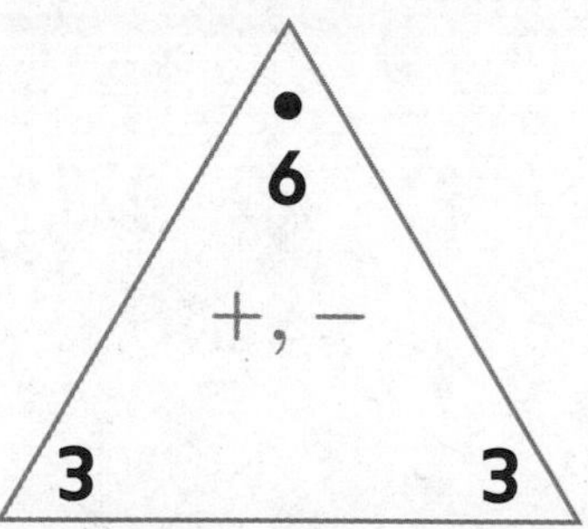

$3 + 3 =$ ______

$6 - 3 = 3$

3 Fill in the missing numbers.

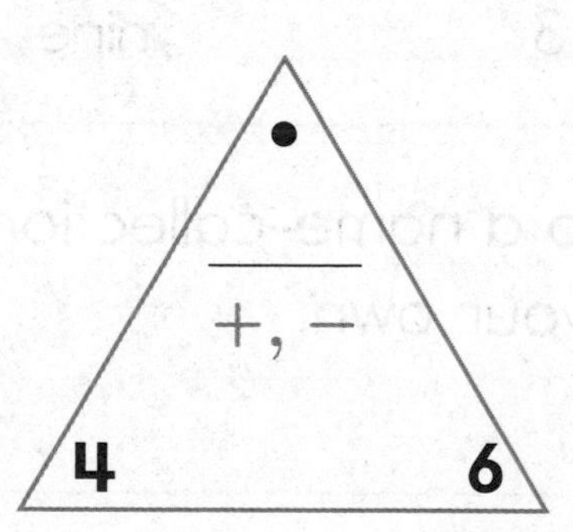

$6 + 4 =$ ______ $10 - 4 = 6$

$4 + 6 =$ ______ $10 - 6 = 4$

4 Fill in the missing numbers.

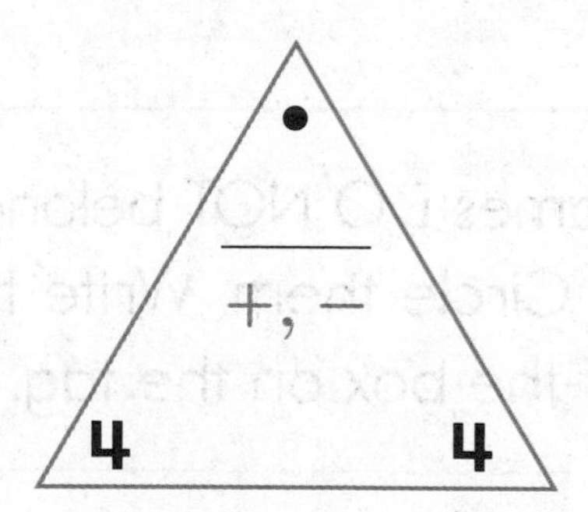

$4 + 4 =$ ______

$8 - 4 = 4$

5 Fill in the empty frames.

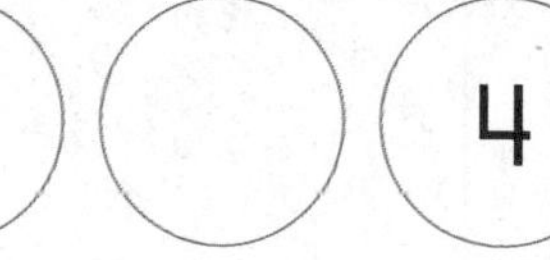

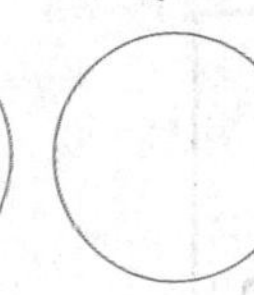

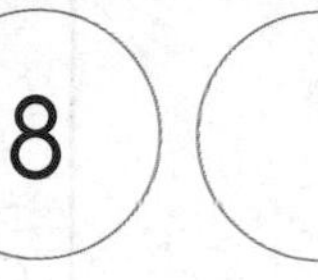

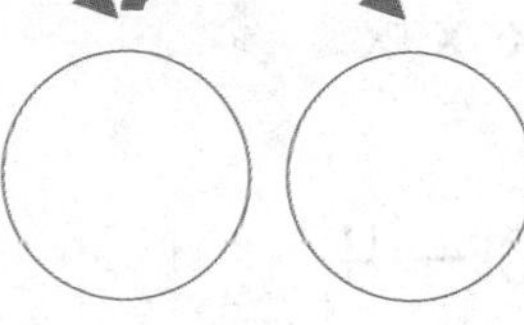

Name-Collection Boxes

Lesson 2-10

DATE

1. Write ten names in the 12 box.

12

2. Circle the names that DO NOT belong in the 9 box.

9

12 − 3

8 + 0

9 − 0

5 + 4 + 1

19 − 10

𝍸 |||

15 − 7

x x x
x x x
x x x

1 less than 10

3 + 3 + 3

nine

3. Three names DO NOT belong in this box. Circle them. Write the name of the box on the tag.

9 + 3

12 − 8

3 + 3

𝍸 ||

x x x
x x x

5 + 3 − 2

10 − 4

half a dozen

4. Make up a name-collection box of your own.

Math Boxes

1. Use your calculator to show 25. The [5] key is broken. Show four ways:

   ______

   ______

   ______

   ______

2. Count back by 5s.

   45, 40, ______, ______, ______

   MRB 66

3. Which number comes next in the pattern?
   48, 58, 68, 78, ______

   (A) 80

   (B) 98

   (C) 75

   (D) 88

   MRB 67

4. Write the number word for 7.

   ______

5. **Writing/Reasoning** Look at Problem 1. How did you show 25 on the calculator without using the 5 key? Explain how you knew what to enter. Then explain why it works.

   ______

   ______

   ______

Adding and Subtracting on a Number Line

Lesson 2-11

DATE

Use the number line to help you solve the number stories.

Show your work.

1. Mihir is collecting rocks. He had 6 rocks. Then he found 4 more. How many rocks does he have in all? ______ rocks

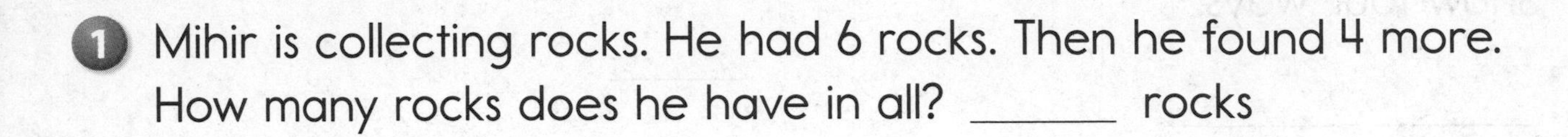

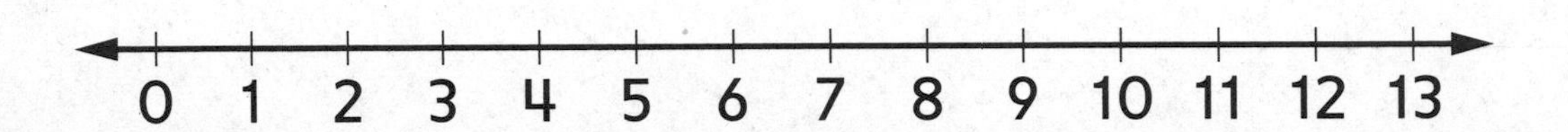

2. Jude had 12 paper airplanes. Then 3 got stuck in a tree. How many does he have left? ______ paper airplanes

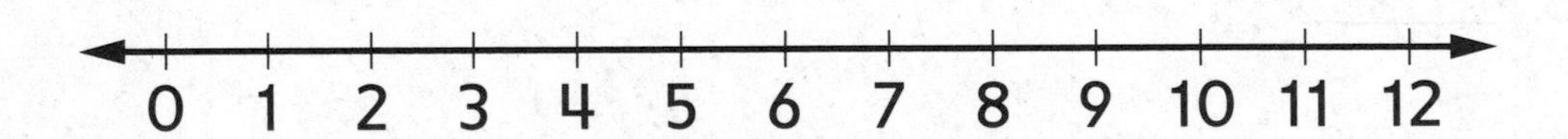

3. Sonu swam 5 laps. Peggy swam 8. How many laps did they swim all together? ______ laps

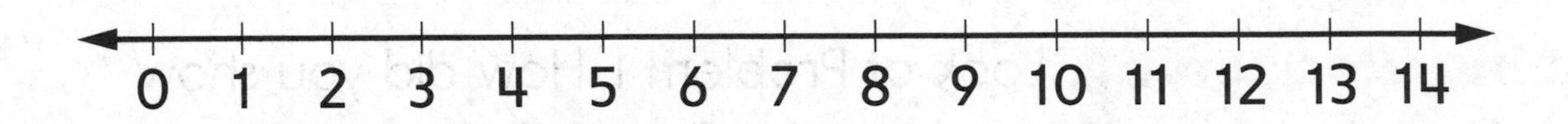

4. Mieke picked a flower with 18 petals. She pulled 9 petals off. How many petals were left? ______ petals

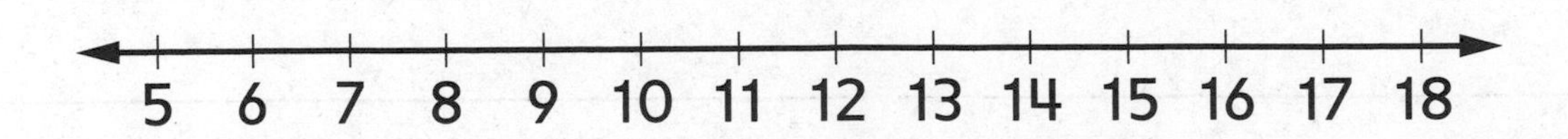

Math Boxes

DATE

1. Fill in the missing numbers.

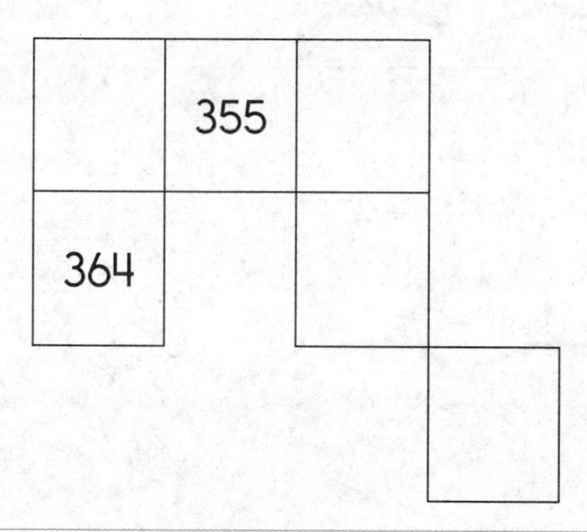

2. Write <, >, or =.

59 ______ 81

194 ______ 149

78 ______ 87

111 ______ 11

< is less than
> is more than
= is the same as

MRB 74-75

3. Fill in the missing numbers.

______, 212, ______, ______,

______, 216, ______

4. Show two ways to make 50¢. Use Ⓟ, Ⓝ, Ⓓ, and Ⓠ.

5. **Writing/Reasoning** Look at Problem 1. Explain how you found the number in the last box.

____

____

____

____

Frames-and-Arrows Problems

Lesson 2-12

DATE

1. Fill in the empty frames.

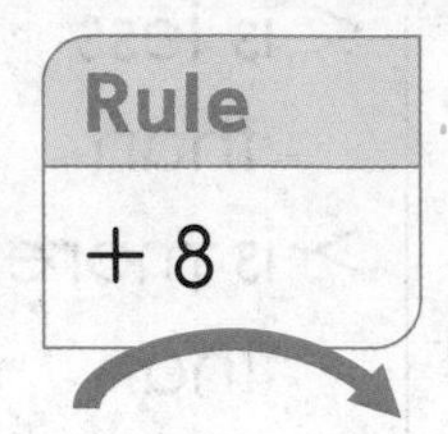

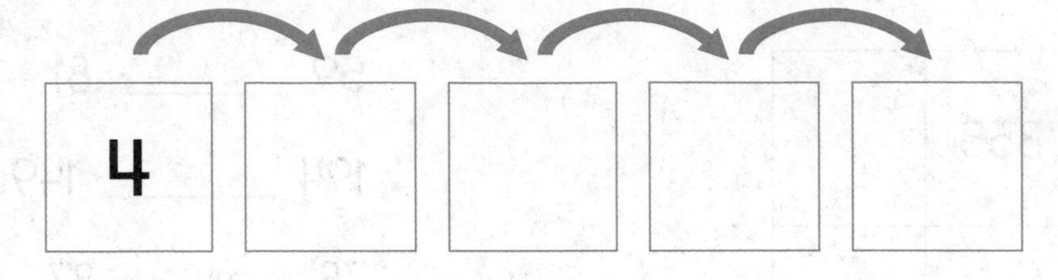

2. Fill in the empty frames.

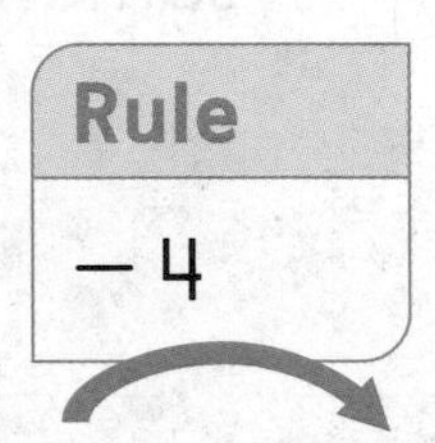

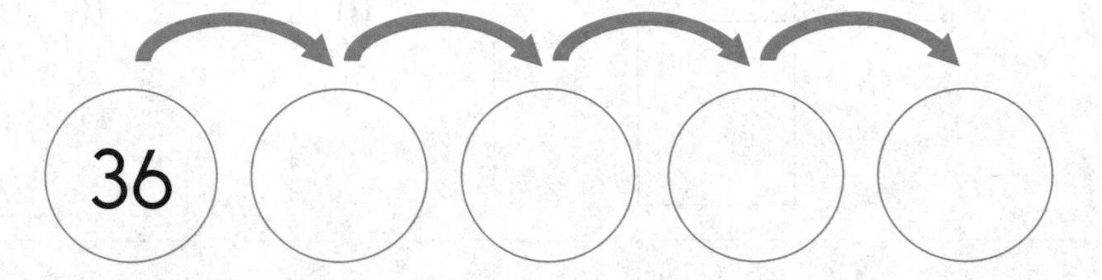

3. Fill in the empty frames.

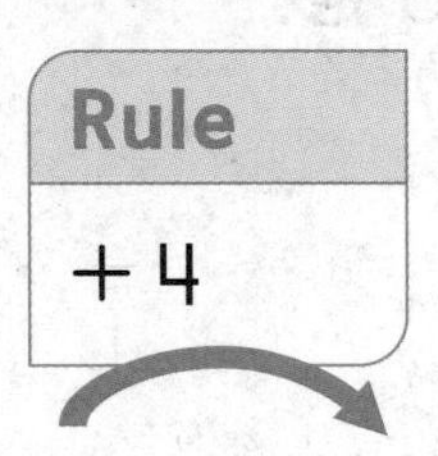

4. Fill in the arrow rule.

Try This

5. Fill in the arrow rule and the empty frames.

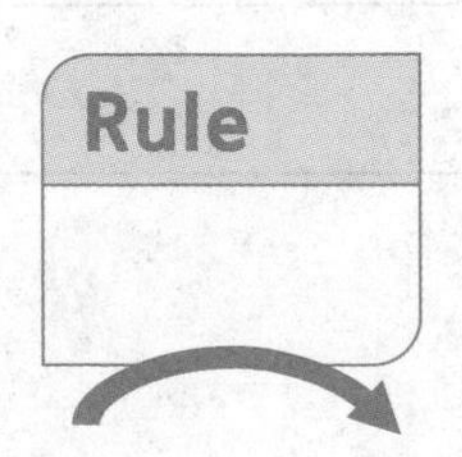

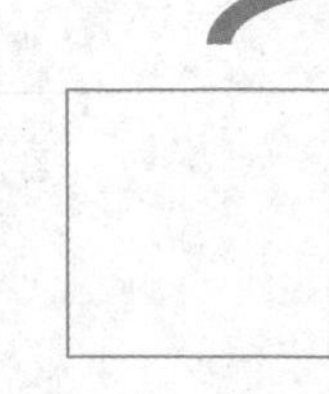

Math Boxes

DATE

1. Use your calculator to show 31.

   The [3] key is broken.

   Show four ways:

   ____

   ____

   ____

   ____

2. Count back by 5s.

   75, 70, ______, ______, 55

3. Which number comes next? Fill in the correct answer.

   150, 160, 170, 180, ______

   (A) 190

   (B) 185

   (C) 200

   (D) 181

4. Write the number word for 5.

   ____

5. **Writing/Reasoning** Write the next five numbers for Problem 3. How did you know what numbers to write?

   ____

   ____

   ____

Math Boxes
Preview for Unit 3

Lesson 2-13

DATE

1. Write the fact family for this Fact Triangle.

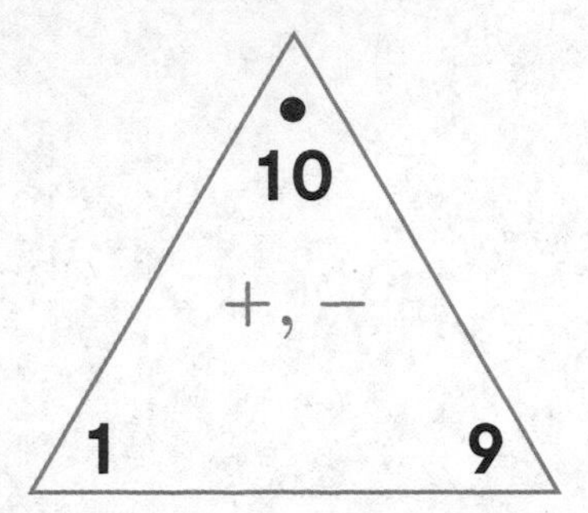

$9 + 1 =$ ______ $10 - 1 = 9$

$1 + 9 =$ ______ $10 - 9 = 1$

2. Write the fact family for this Fact Triangle.

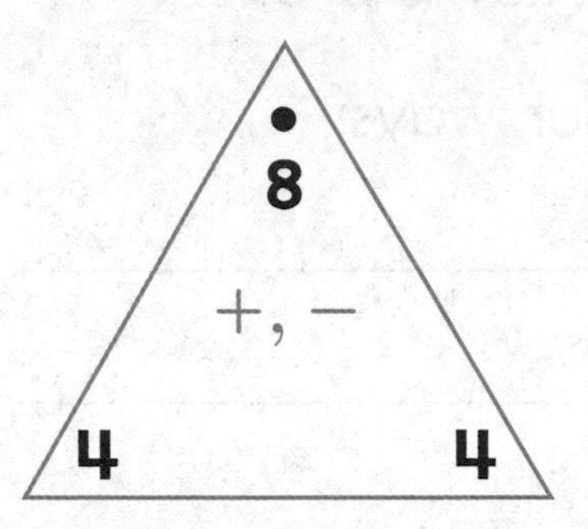

$4 + 4 =$ ______ $8 - 4 = 4$

3. Fill in the missing numbers.

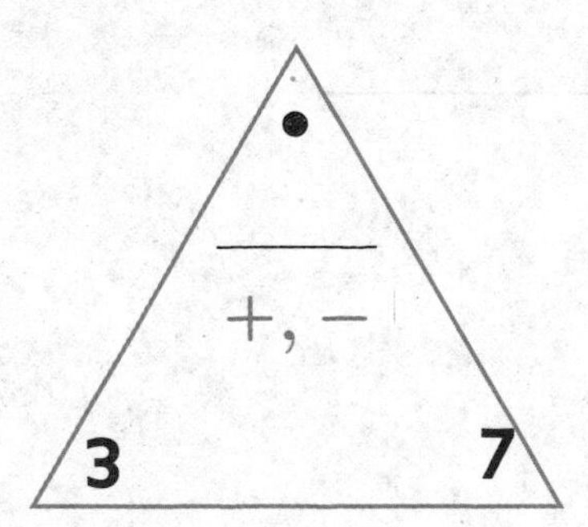

$7 + 3 =$ ______ $10 - 7 = 3$

$3 + 7 =$ ______ $10 - 3 = 7$

4. Fill in the missing numbers.

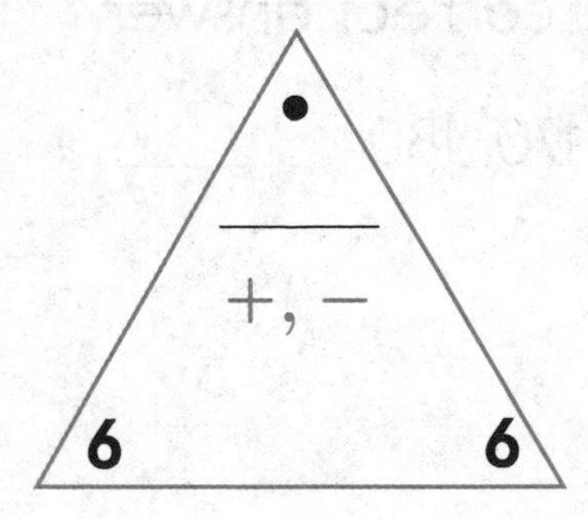

$6 + 6 =$ ______ $12 - 6 = 6$

5. Fill in the empty frames.

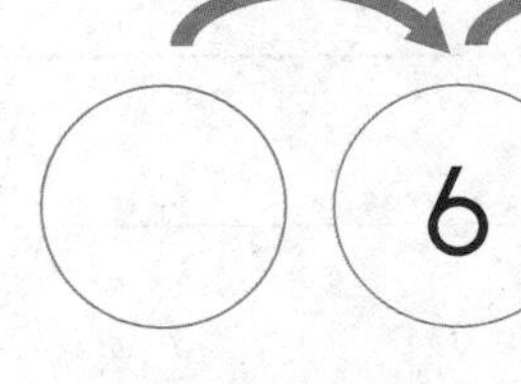

Making 10 on a Double Ten Frame

Lesson 3-1

DATE

Make sure you and your partner have 20 counters and a blank double ten frame. Your teacher will show you a picture of dots on a double ten frame.

1. Figure out how many dots there are on the double ten frame without counting them one by one.
2. Explain to your partner how you solved the problem. Use counters and a double ten frame to help show your thinking.

Math Boxes

1 Write the number word for 4.

____

2 Use your calculator to show 19.
The (9) key is broken.
Show four ways:

____

____

____

____

3 Some oranges were in a bowl. Kalani added 3 more. Now there are 6. How many oranges were in the bowl before Kalani added 3 more? ______

Unit
oranges

Number model:

____

MRB 30-31

4 Fill in the missing numbers.

______, 129, ______, ______,

132, ______, ______

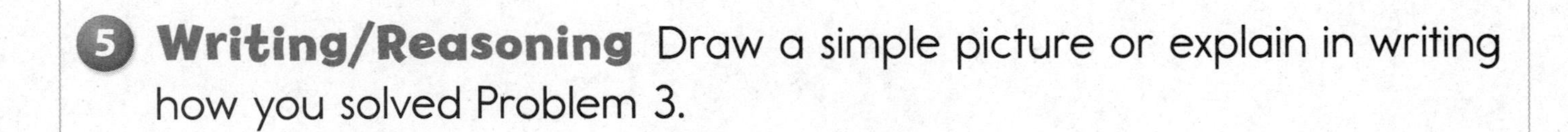

5 **Writing/Reasoning** Draw a simple picture or explain in writing how you solved Problem 3.

MRB 30-31

Domino Facts

Lesson 3-2

DATE

Write addition and subtraction facts for each domino.

1

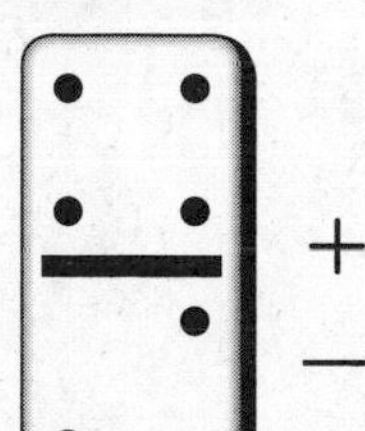

$$4 + 2 = 6 \quad 2 + 4 = 6 \quad 6 - 2 = 4 \quad 6 - 4 = 2$$

2

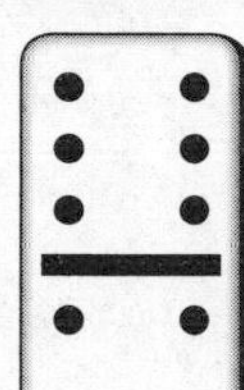
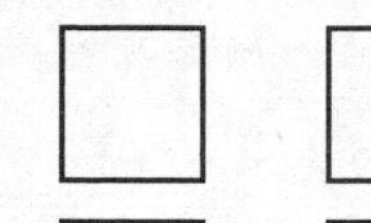
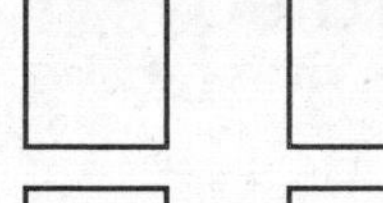

$\square + \square$ $\square + \square$ $\square - \square$ $\square - \square$

3

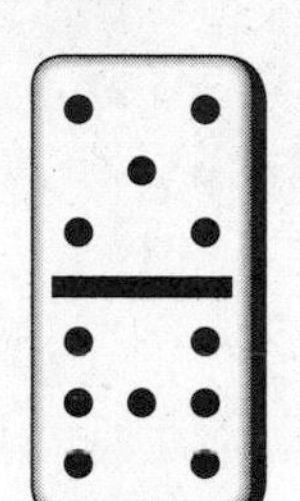
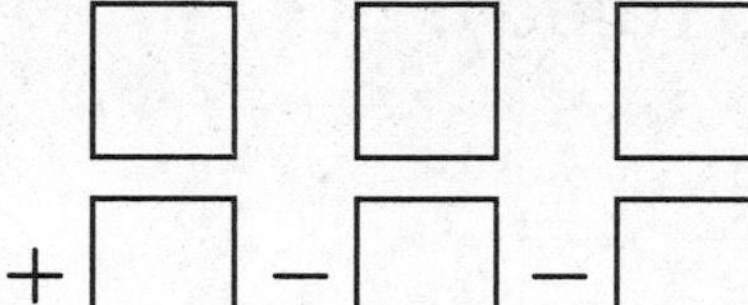

$\square + \square$ $\square + \square$ $\square - \square$ $\square - \square$

4

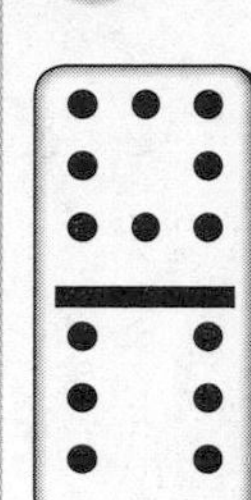
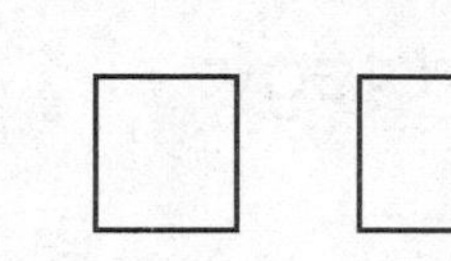
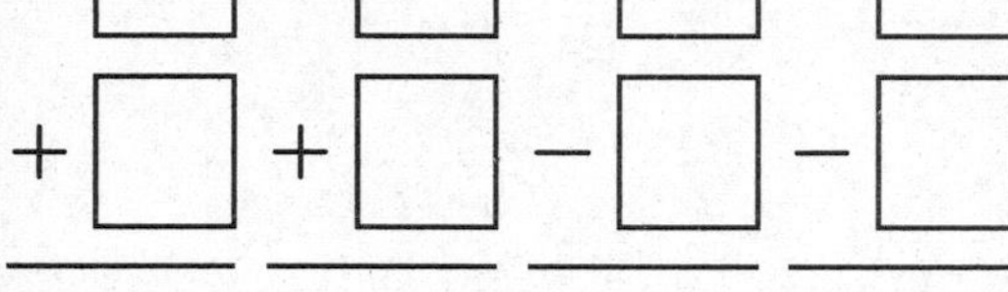

$\square + \square$ $\square + \square$ $\square - \square$ $\square - \square$

5

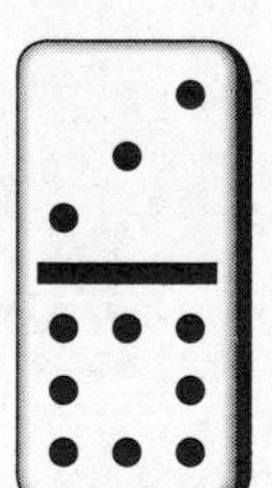
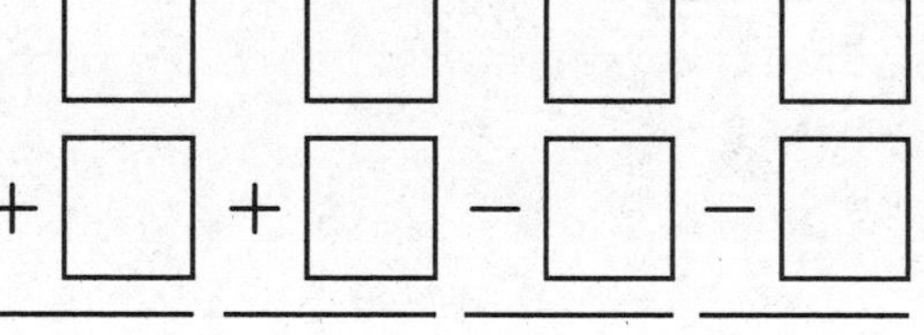

$\square + \square$ $\square + \square$ $\square - \square$ $\square - \square$

6
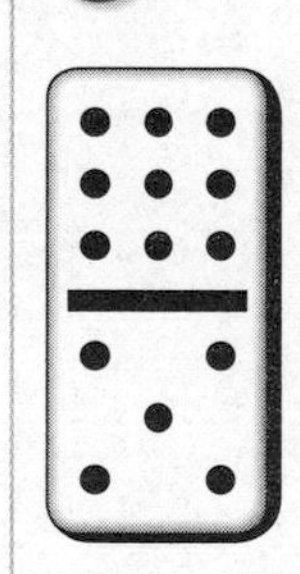
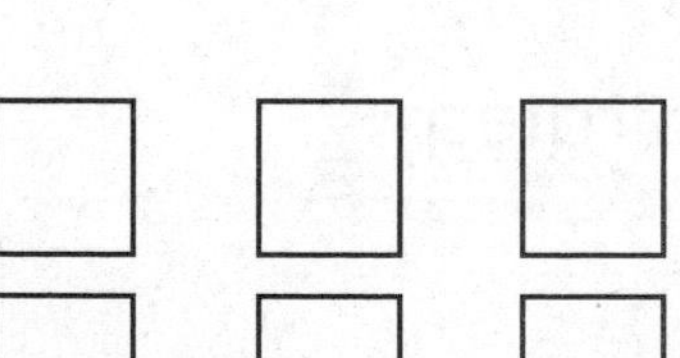

$\square + \square$ $\square + \square$ $\square - \square$ $\square - \square$

7

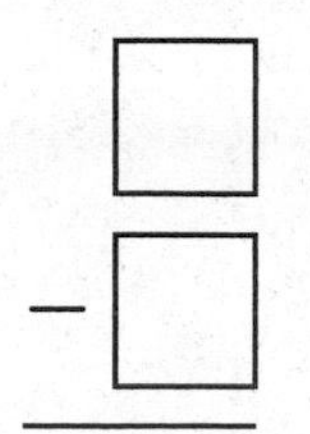

$\square + \square$ $\square - \square$

Try This

8

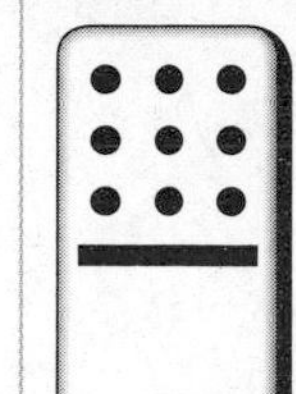
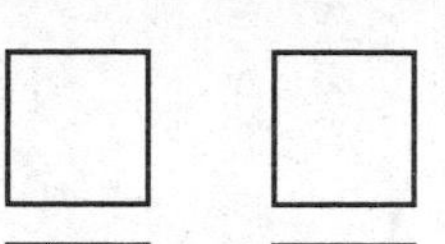
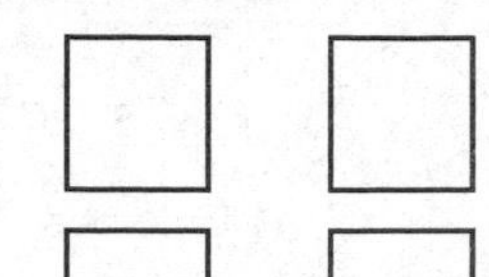

$\square + \square$ $\square + \square$ $\square - \square$ $\square - \square$

My Subtraction Fact Strategies

Lesson 3-2

DATE

Use this page to record subtraction strategies. When you don't know how to solve a subtraction problem, you can look here for ideas.

Strategy:	Strategy:
Strategy:	Strategy:
Strategy:	Strategy:
Strategy:	Strategy:

Math Boxes

DATE

1 Write <, >, or =.

9 + 7 ______ 13

10 + 10 ______ 26

7 + 7 ______ 5 + 9

MRB 75

2 How many cents are there?

Ⓟ Ⓝ Ⓓ Ⓓ Ⓠ Ⓠ

Fill in the circle next to the best answer.

Ⓐ 74¢
Ⓑ 81¢
Ⓒ 76¢
Ⓓ 51¢

MRB 110-111

3 ● ● ● ●
● ● ● ●
● ● ● ●
● ● ● ●
● ● ● ●

How many dots? ______

Odd or even? ______________

MRB 59-60

4 Write six names for 15.

15

MRB 53

5 Fill in the empty frames.

Rule
+ 2

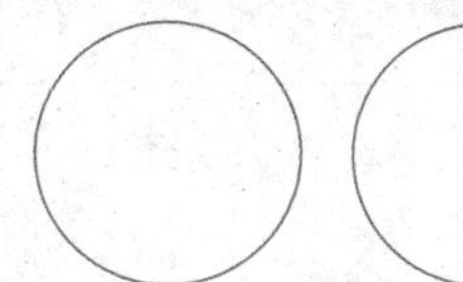
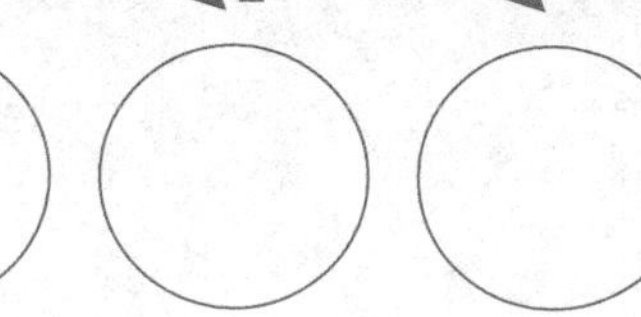

( ) → (5) → ( ) → ( ) → (11) → ( ) → ( )

MRB 54-55

Addition/Subtraction Facts Table

Lesson 3-3

DATE

+, −	0	1	2	3	4	5	6	7	8	9
0	0	1	2	3	4	5	6	7	8	9
1	1	2	3	4	5	6	7	8	9	10
2	2	3	4	5	6	7	8	9	10	11
3	3	4	5	6	7	8	9	10	11	12
4	4	5	6	7	8	9	10	11	12	13
5	5	6	7	8	9	10	11	12	13	14
6	6	7	8	9	10	11	12	13	14	15
7	7	8	9	10	11	12	13	14	15	16
8	8	9	10	11	12	13	14	15	16	17
9	9	10	11	12	13	14	15	16	17	18

Math Boxes

Lesson 3-3

DATE

Math Boxes

1. Write the number word for 2.

____

2. Use your calculator to show 23.

The (3) key is broken. Show four ways:

____

____

____

____

3. Some dogs are at the dog park. Five more dogs join them. Now there are 10. How many dogs were at the park to start?

Answer: ______

Number model:

____

Unit
dogs

4. Fill in the missing numbers.

199, ______, ______, 202,

______, ______, 205

5. **Writing/Reasoning** Look at Problem 2. Explain how you found one way to show 23.

____

____

____

Many Names for Numbers

Lesson 3-4

DATE

1. Write ten names in the 12 box.

2. Circle the names that do NOT belong in the 10 box.

10

5 + 6	
7 + 3	ten
8 + 2	20 − 10
𝍸 𝍸 /	11 − 2

3. Three names do NOT belong in this box. Circle them. Write the name of the box on the tag.

4 + 3	
𝍸 //	6 + 1
5 + 3	2 + 2 + 3
9 − 3	8 − 1 − 1
10 − 3	

4. Make your own name-collection box.

Math Boxes

1 Write <, >, or =.

$6 + 7$ ______ $15 - 4$

$5 + 8$ ______ $8 + 5$

$18 - 9$ ______ $5 + 5$

MRB 75

2 How many cents are there?

Ⓟ Ⓟ Ⓓ Ⓓ Ⓠ Ⓠ Ⓠ

Fill in the circle next to the best answer.

Ⓐ 97¢
Ⓑ 75¢
Ⓒ 52¢
Ⓓ 67¢

3

How many dots are there in all?

Answer: ______

Odd or even? ______

4 Write the label for the name-collection box. Add three more names to the box.

$17 - 1$ $18 - 2$ sixteen

5 Fill in the empty frames.

Rule
+ 10

20		40			70

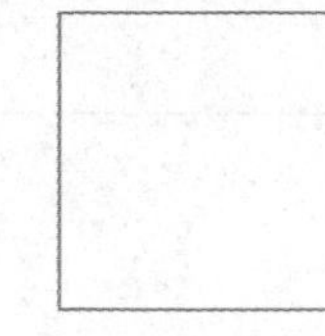

MRB 54-55

Math Boxes

Lesson 3-5

DATE

Math Boxes

1 Count up by 10s.

96, ______, ______, 126,

______, 146, ______

2 Write the number word for 6.

____

3 Write <, >, or =.

4 + 6 ______ 6 + 3

5 + 9 ______ 7 + 7

8 + 5 ______ 9 + 6

4 Show 75¢ two ways.

5 **Writing/Reasoning** Look at Problem 4. Suppose you want to use as few coins as possible to make 75¢. What coins would you use? Use words or pictures to explain your answer.

____

____

____

MRB 110-111

Math Boxes

DATE

1. Write these numbers in order from smallest to largest.

   23, 59, 49, 3, 159

   ______, ______, ______,

   ______, ______

   MRB 74

2. Count back by 2s from 30.

   Fill in the circle next to the correct answer.

   Ⓐ 30, 26, 22, 18, 14
   Ⓑ 30, 25, 20, 15, 10
   Ⓒ 30, 29, 28, 27, 26
   Ⓓ 30, 28, 26, 24, 22

3. Write six names for 18.

   MRB 53

4. Fill in the missing numbers.

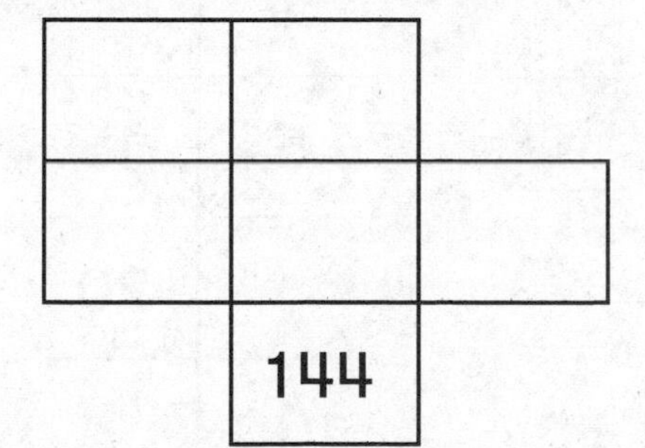

5. Use your calculator to count by 6s. Fill in the empty frames.

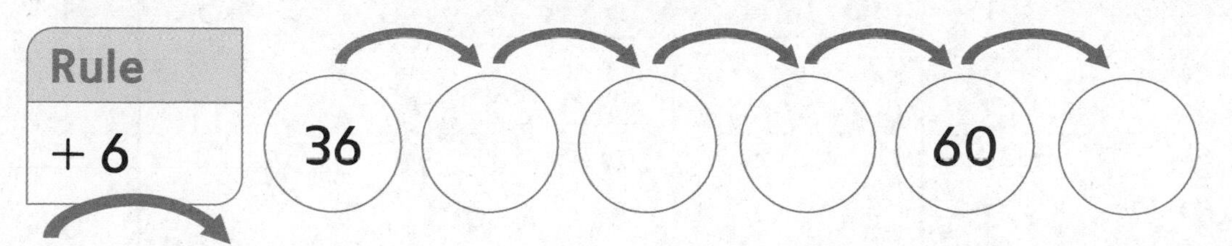

What's My Rule?

Lesson 3-7

DATE

In Problems 1–4, follow the rule. Fill in the missing numbers.

1

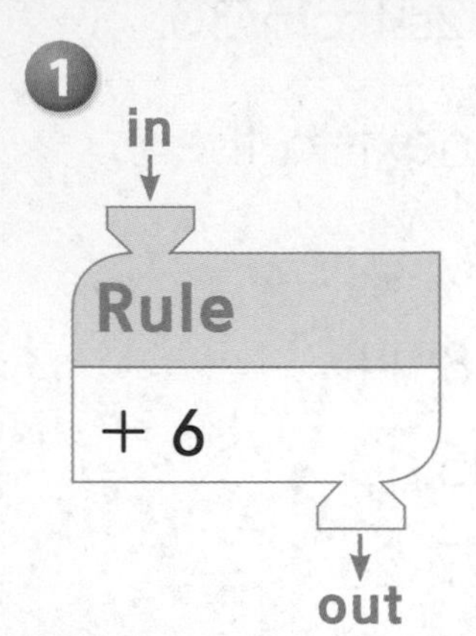

in	out
2	8
3	9
5	
9	

2

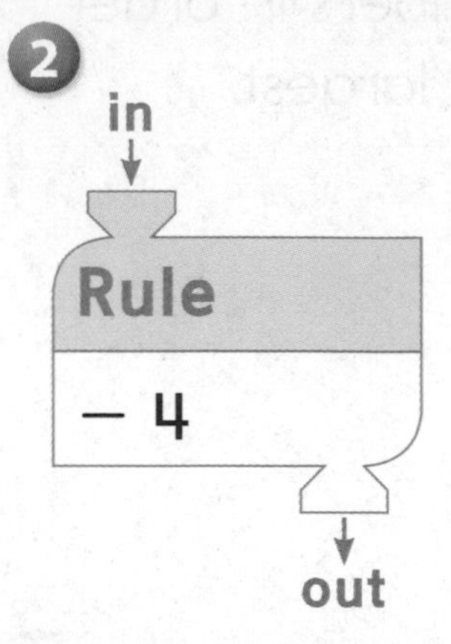

in	out
6	2
8	
10	
5	

3

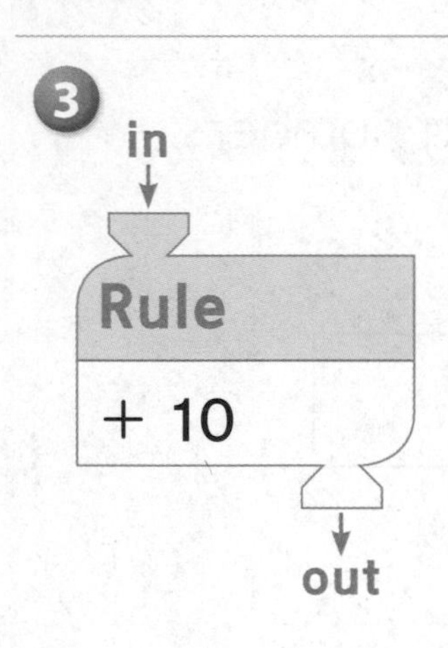

in	out
1	
5	15
	20
100	

4

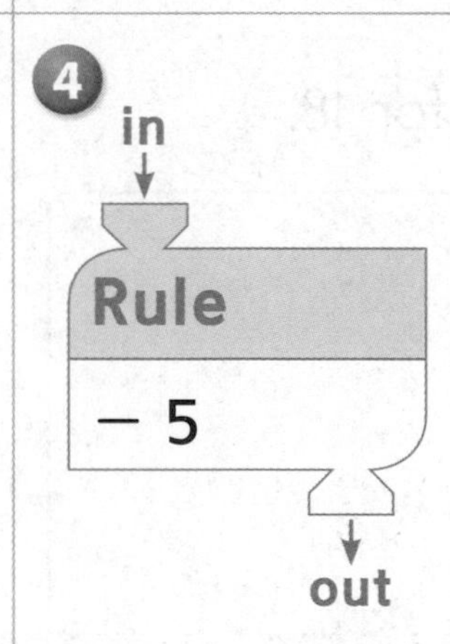

in	out
6	
	3
5	
12	

In Problems 5–6, find the rule. Write it in the box.
Then fill in any missing numbers.

5

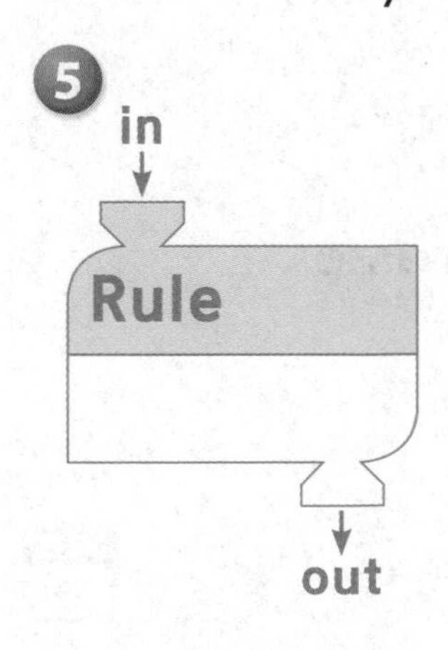

in	out
6	13
1	8
3	
4	

6

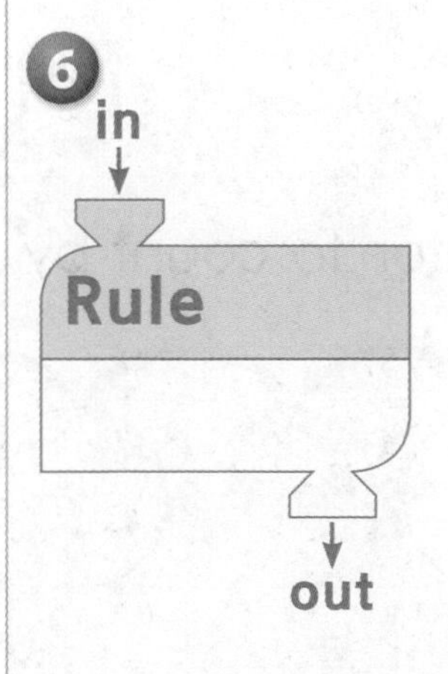

in	out
12	10
6	4
11	
	6

Math Boxes

DATE

1 Count back by 10s.

113, ______, ______, 83, ______,

______, 53

2 Write the number word for 8.

____

3 Write $<$, $>$, or $=$.

$4 + 4$ ______ $5 + 4$

$7 + 7$ ______ $8 + 6$

$9 + 8$ ______ $8 + 8$

MRB 75

4 Use coins to show $1 two ways.

5 **Writing/Reasoning** In Problem 1, count back by 10s five more times. What patterns do you see?

____

____

____

MRB 67

Using Doubles to Subtract

DATE

1 Solve the doubles fact. Then use it to help you solve the two subtraction facts.

a.	b.
4 + 4 = ______	9 + 9 = ______
8 − 4 = ______	18 − 9 = ______
9 − 4 = ______	17 − 9 = ______

2 Think of a doubles fact that can help you solve each problem. Record your helper fact and then solve the problem.

a.	b.
9 − 5 = ?	13 − 7 = ?
Helper fact:	Helper fact:
______________________	______________________
9 − 5 = ______	13 − 7 = ______

3 Explain how to solve 15 − 8 = ? using a doubles helper fact.

____

____

____

Math Boxes Preview for Unit 4

Lesson 3-8

DATE

1. Write the time.

______

2. Write <, > or =.

46 ______ 36

33 ______ 43

65 ______ 89

3. What number is shown by the base-10 blocks?

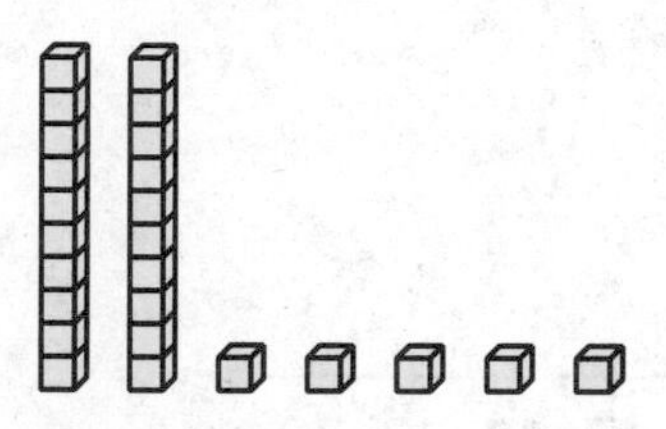

Ⓐ 23
Ⓑ 25
Ⓒ 52
Ⓓ 205

MRB 73

4. Circle ten stars.

How many groups of 10 are there? ______

How many stars are left over?

______

How many stars in all? ______

5. How many base-10 cubes long is the marker? ______ cubes

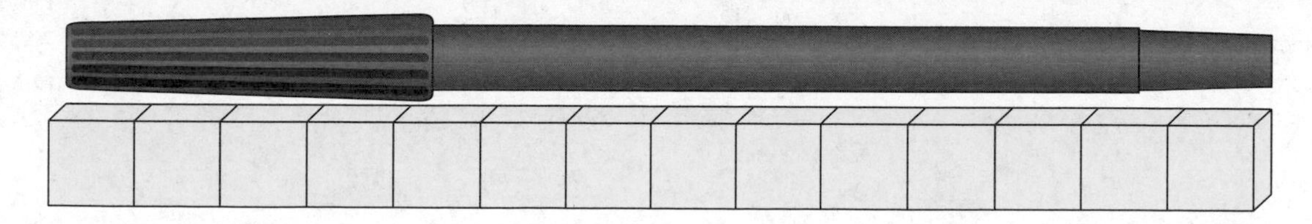

The Going-Back-Through-10 Strategy

Lesson 3-9

DATE

Use the going-back-through-10 strategy to solve each fact. Show your work on the number line. Then explain your work to a partner.

Example: 15 − 6 = 9

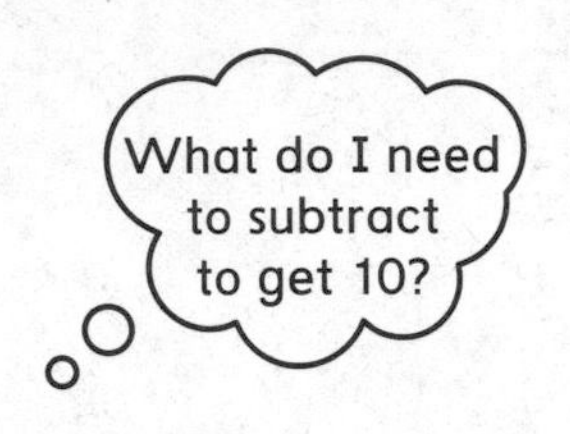

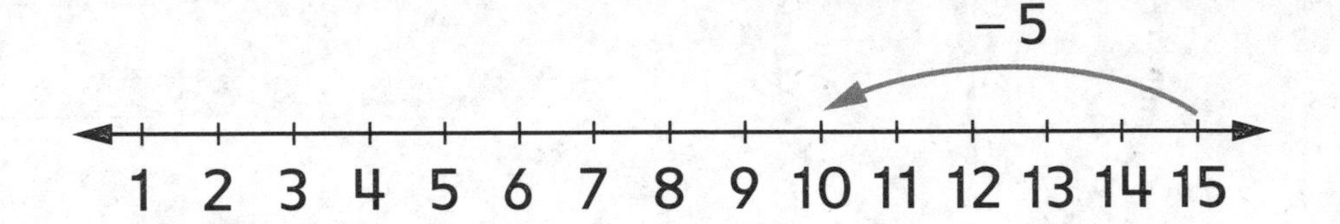

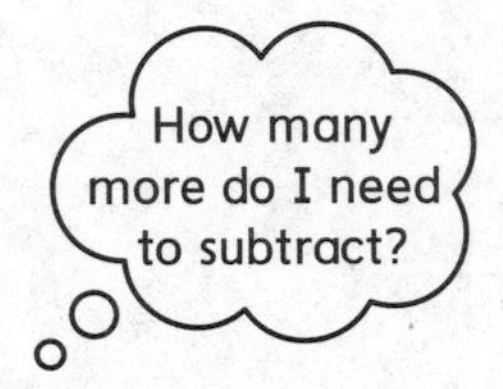

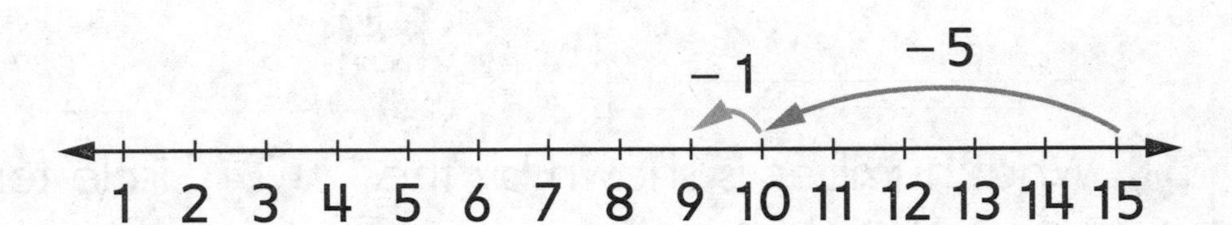

1. 14 − 5 = ______

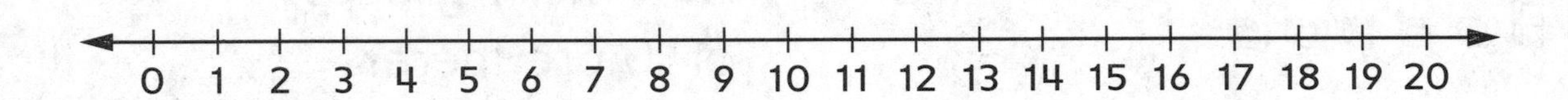

2. 13 − 8 = ______

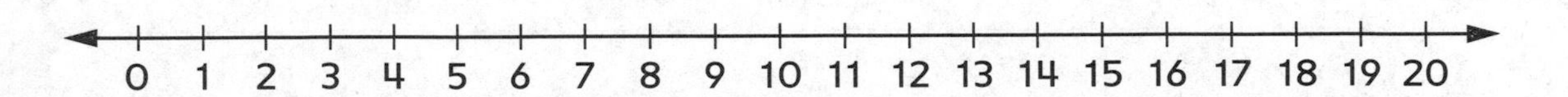

3. 17 − 8 = ______

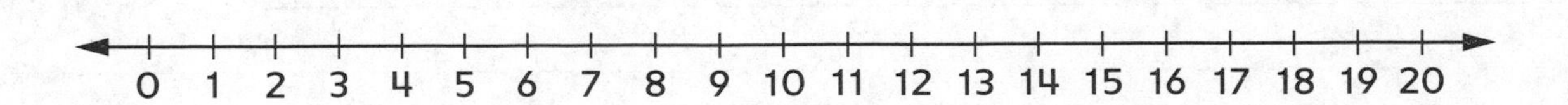

Subtraction Number Stories

Lesson 3-9

DATE

Solve each problem. Look at your My Subtraction Fact Strategies list on journal page 48 to help you. Show your work.

1. Dajon has $11. He buys a book for $6. How much money does he have left?

   Dajon has $ ______ left.

2. Julia has 7 markers. Carlos has 4 markers. How many more markers does Julia have than Carlos?

   Julia has ______ more markers than Carlos.

3. Martin had some flowers. He gave 4 flowers to his sister and had 6 left. How many flowers did Martin have to start?

   Martin had ______ flowers to start.

4. Make up and solve your own subtraction story.

   ____

   ____

   ____

   ____

   ____

   ____

Math Boxes

Lesson 3-9

DATE

Math Boxes

1. Count on by 5s from 175.

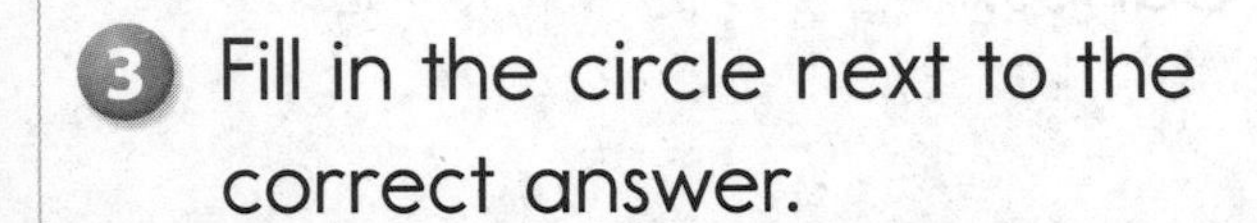

175, ______, ______, ______,

______, ______, ______

2. Write the number word for 10.

______

3. Fill in the circle next to the correct answer.

There were 11 children at the park. Then 7 left. How many children are still at the park?

Ⓐ 5
Ⓑ 3
Ⓒ 4
Ⓓ 6

4. Write the missing numbers.

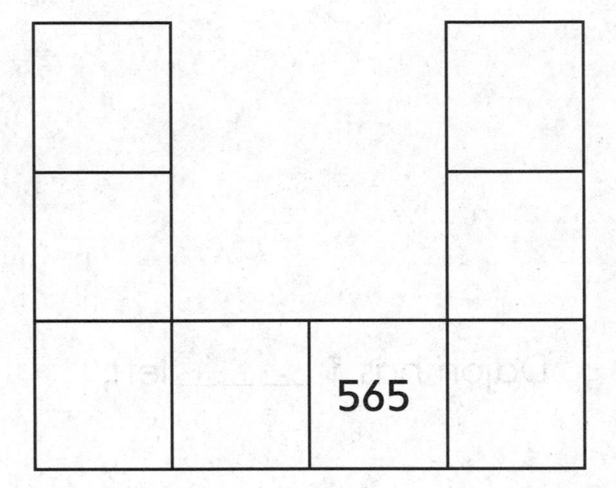

MRB 68

5. **Writing/Reasoning** Look at Problem 4.
Explain how you completed the number-grid puzzle.

______

______

______

Using 10 as a Friendly Number

Lesson 3-10

DATE

Use the number line to solve the problem. Do not count by 1s.

1. Anna's father likes to run. Today he will run 17 miles. He has run 8 miles so far. How many more miles will he run today?

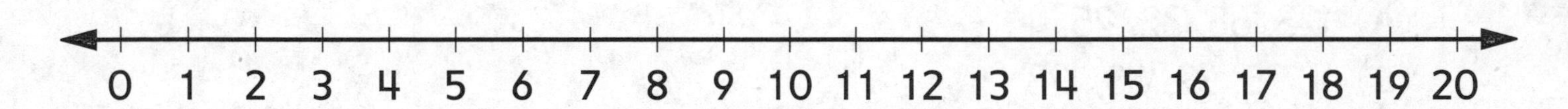

Answer: ______ miles

For Problems 2–4, use the number lines to show how to solve each problem using the "friendly number" 10. Explain how you solved the problems to a partner.

Unit
miles

2. $15 - 7 =$ ______

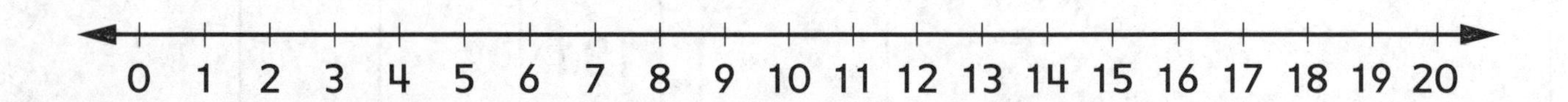

3. $16 - 9 =$ ______

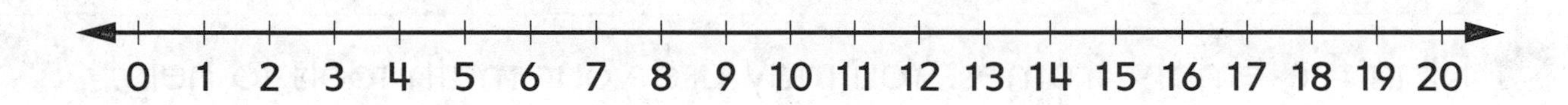

4. $14 - 5 =$ ______

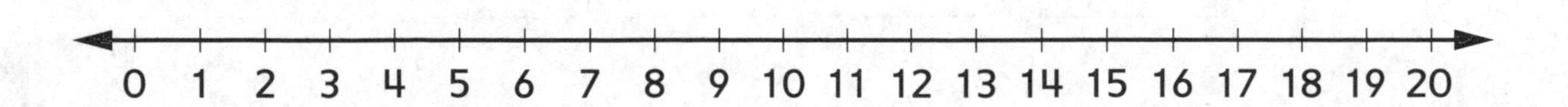

Math Boxes

Lesson 3-10

DATE

1. Which set of numbers is ordered from smallest to largest? Fill in the circle next to the correct answer.

(A) 123, 68, 32, 25, 4
(B) 4, 25, 32, 68, 123
(C) 4, 32, 25, 68, 123
(D) 32, 4, 123, 25, 68

MRB 74

2. Count on by 10s.

360, _____, _____, _____, _____, 410, _____

MRB X-X

3. Write six names for 25.

25

MRB 53

4. Fill in the missing numbers.

	122	
131		

5. Fill in the empty frames. You may use your math tools to help.

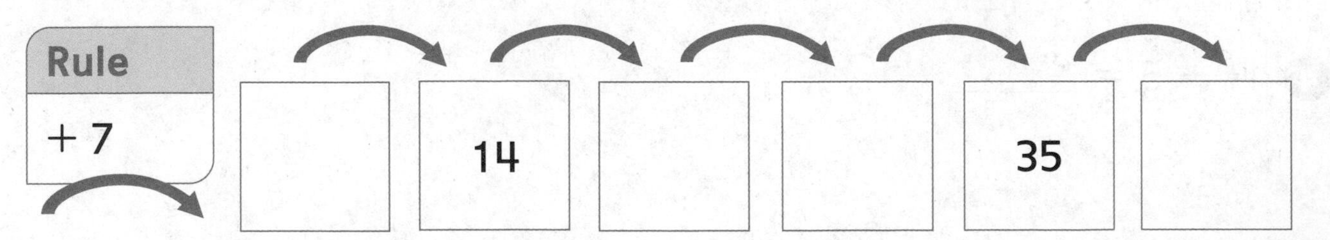

MRB 54-55

Math Boxes

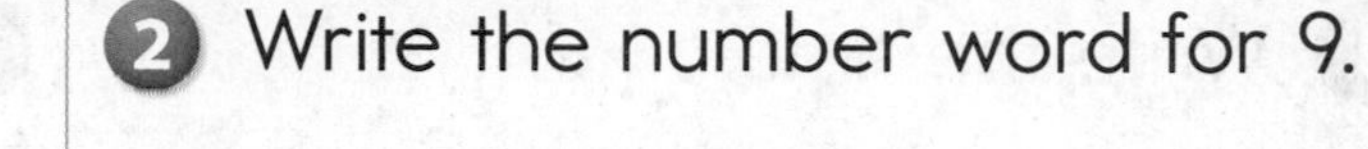

1. Count back by 10s. Fill in the circle next to the correct answer.

   Ⓐ 810, 820, 830, 840, 850
   Ⓑ 810, 815, 820, 825, 830
   Ⓒ 810, 800, 790, 780, 770
   Ⓓ 810, 710, 610, 510, 410

2. Write the number word for 9.

   ____

3. Some fish were in a tank. After 9 fish were added, there were 12 in all. How many fish were in the tank at the start?

Unit
fish

   _____ fish

   Number model:

   ____

4. Write the missing numbers.

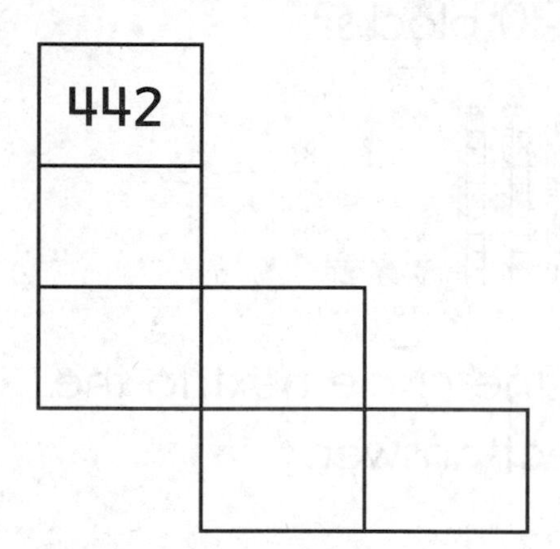

5. **Writing/Reasoning** Use pictures or words to explain how you solved Problem 3.

____

____

Math Boxes
Preview for Unit 4

Lesson 3-12

DATE

Math Boxes

1 Write the time.

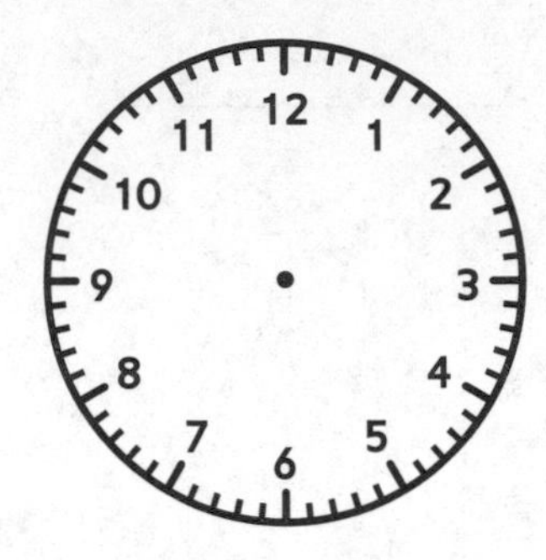

______

MRB 106-107

2 Write <, > or =.

31 ______ 21

43 ______ 51

93 ______ 83

MRB 74-75

3 What number is shown by the base-10 blocks?

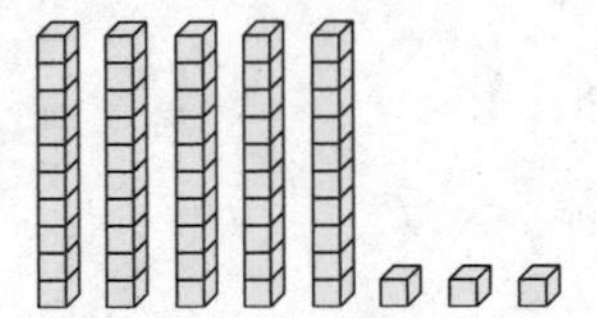

Fill in the circle next to the correct answer.

Ⓐ 23

Ⓑ 35

Ⓒ 53

Ⓓ 503

4 Circle 10 cubes.

How many groups of ten are there? ______

How many are left over?

______

How many cubes are there in all? ______

5 About how many base-10 cubes long is the crayon?

______ cubes

What Time Is It?

Lesson 4-1

DATE

Write the time.

1  ___ : ___	2  ___ : ___ What time will it be in 1 hour? ___ : ___	3  ___ : ___ What time will it be in 2 hours? ___ : ___

Draw clock hands to match each time.

4 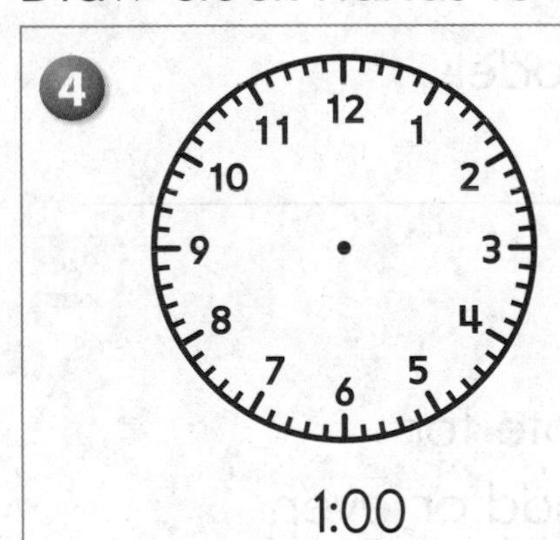 1:00	5 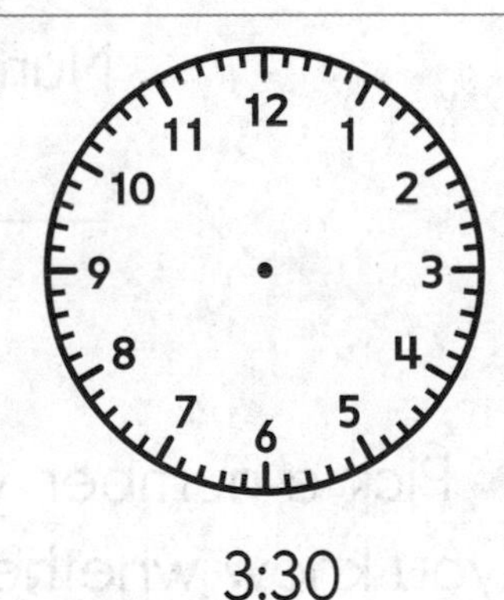 3:30	6 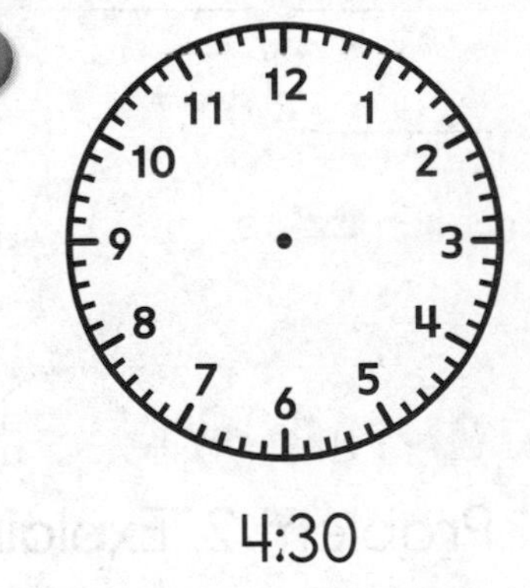 4:30

Try This

7 My family went to a movie. It started at 7:30 at night. If ended at 9:30 at night. How long was the movie? ___

Explain how you found your answer. ___

___

___

Math Boxes

DATE

1. Write the number word for 11.

   ______

2. Write three 2-digit numbers that are even. Then write three 2-digit numbers that are odd. Circle the even numbers.

   ______ ______ ______

   ______ ______ ______

   MRB 60

3. Write the numbers in order from smallest to largest.

   11, 32, 70, 4, 25

   ______, ______,

   ______, ______,

   ______

   MRB 74

4. Chin has 7 more cards than Li. Li has 4 cards. How many cards does Chin have?

Unit
cards

   ______ cards

   Number model:

   ______

   MRB 30–31

5. **Writing/Reasoning** Pick a number you wrote for Problem 2. Explain how you know whether it is odd or even.

   ______

   ______

   ______

What Time Is It?

Lesson 4-2

DATE

Write the time.

1

___:___

2

___:___

3

___:___

Draw the hands to match the time.

4
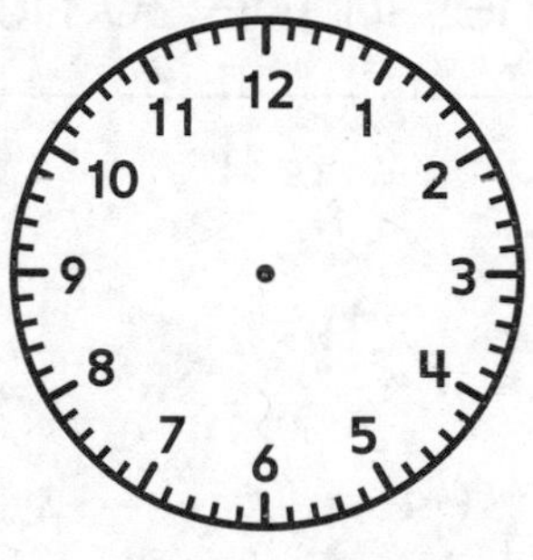

5:30

5
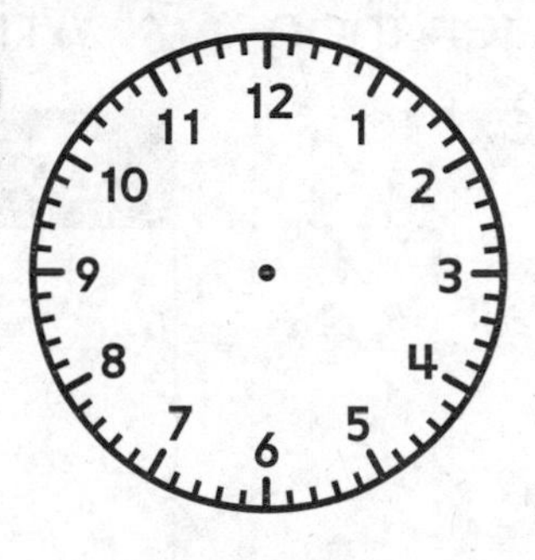

1:45

6
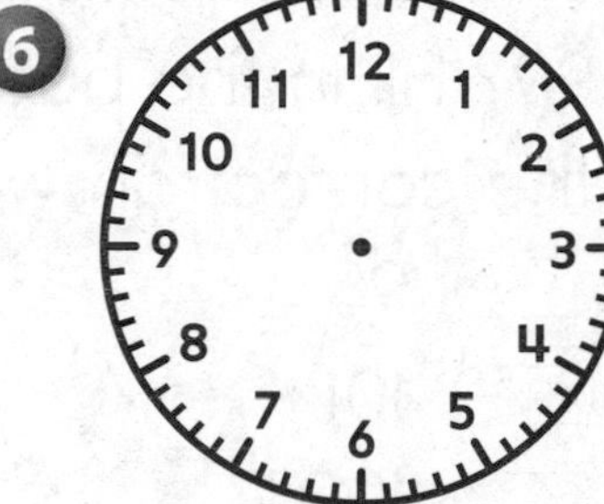

4:05

7 Make up your own. Draw the hands to show each time.
Write the time below each clock.

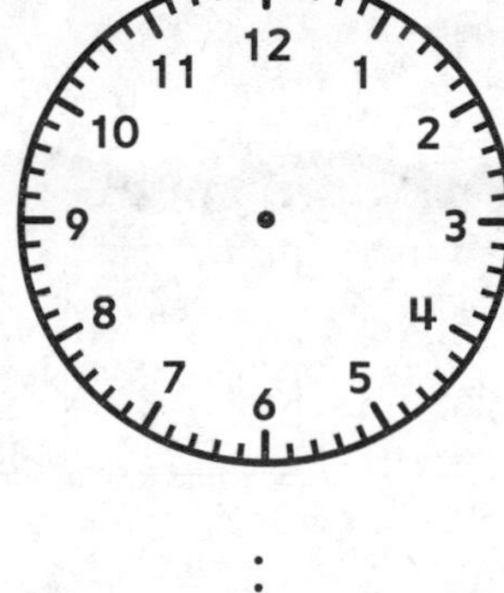

___:___

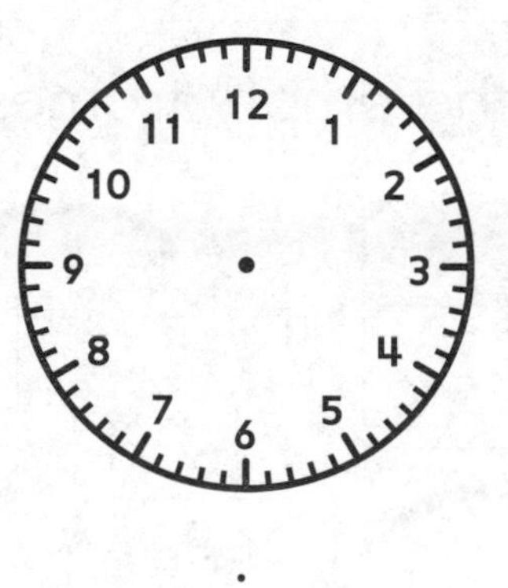

___:___

Math Boxes

Lesson 4-2

DATE

Math Boxes

1. Find the amount. Fill in the bubble next to the correct answer.

Ⓝ Ⓝ Ⓠ Ⓟ Ⓓ Ⓓ

- ◯ 56¢
- ◯ 55¢
- ◯ 30¢
- ◯ 46¢

MRB 110–111

2. Fill in the missing numbers.

Rule
+ 9

in	out
10	19
3	12
	13
6	

MRB 56–57

3. Which numbers are larger than 99? Fill in the bubbles next to the correct answers.

- ◯ 98
- ◯ 101
- ◯ 100
- ◯ 90

MRB 74

4. Write six names in the 20 box.

MRB 53

5. Fill in the empty frames.

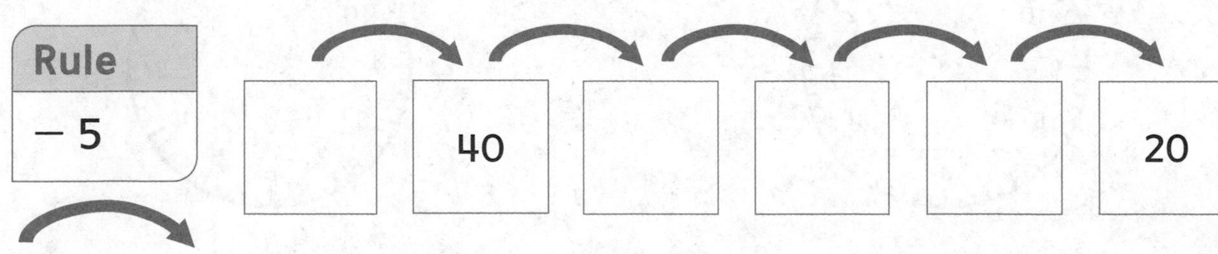

MRB 54–55

DATE

A.M. and P.M.

Draw a picture to show something that usually happens at each time below.

8:00 A.M.	8:00 P.M.
11:00 A.M.	11:00 P.M.

Math Boxes

1. Write the number word for 13.

   ____

2. Write three 2-digit numbers that are even. Then write three 2-digit numbers that are odd. Circle the odd numbers.

   ______ ______ ______

   ______ ______ ______

   MRB 60

3. Write the numbers in order from largest to smallest.

   36, 21, 63, 8, 12

   ______, ______, ______,

   ______, ______,

   MRB 74

4. Rachel has 8 books. This is 3 more than Amy. How many books does Amy have?

Unit
books

   ______ books

   Number model:

   ____

   MRB 30–31

5. **Writing/Reasoning** Explain how you know which is the largest number in Problem 3.

   ____

   ____

   ____

Math Boxes

Lesson 4-4

DATE

1 How much money is shown?

Ⓟ Ⓟ Ⓝ Ⓝ Ⓝ Ⓓ Ⓠ Ⓠ

_______¢

MRB 110–111

2 Fill in the rule and the missing numbers.

Rule

in	out
42	52
11	21
	108
74	

MRB 56–57

3 Which numbers are larger than 750? Fill in the bubbles next to the correct answers.

- ◯ 749
- ◯ 801
- ◯ 751
- ◯ 738

MRB 74

4 Write six names for 50.

50

MRB 53

5 Fill in the empty frames.

Rule

− 7

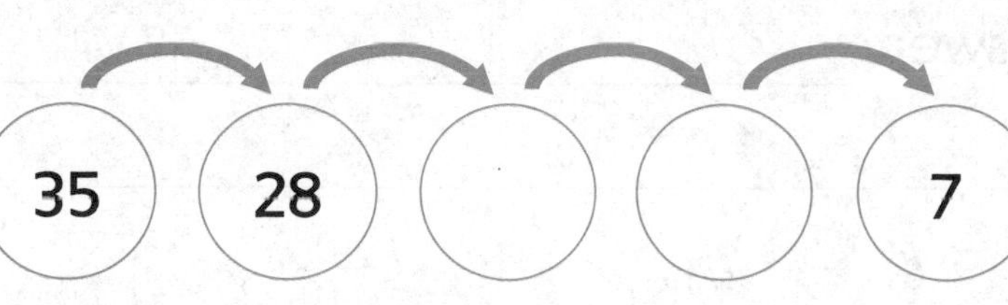

MRB 54-55

Comparing Numbers

DATE

Write each number in expanded form. Then write < or > in the box to compare the two numbers.

Example: 82 = 80 + 2

87 = 80 + 7

82 [<] 87

1. 68 = ______

   57 = ______

   68 [ ] 57

2. 42 = ______

   48 = ______

   42 [ ] 48

3. 312 = ______

   321 = ______

   312 [ ] 321

4. 570 = ______

   507 = ______

   570 [ ] 507

Write < or > in the box.

5. 456 [ ] 546

6. 213 [ ] 203

Try This

7. Sara says that 125 and 1,250 have the same value. Do you agree? Explain your answer. ______

Math Boxes

1 Write the time.

_____ : _____

MRB 106

2 Mateo has 15 stickers. That is 7 fewer stickers than Dan. How many stickers does Dan have?

Unit
stickers

Number model:

____

Answer: _____ stickers

MRB 30–31

3 Use a number grid. Find the distance from . . .

15 to 45. _____

25 to 75. _____

4 Fill in the blanks to complete the pattern.

90, _____, _____, 75, 70

5 **Writing/Reasoning** How did you find the distances in Problem 3?

____

____

____

Representing Numbers

Lesson 4-6

DATE

Solve the problems.

Use □ to show a flat, | to show a long, and ▪ to show a cube.

1. Use base-10 blocks to show the number 23. Draw your blocks below.

2. Use base-10 blocks to show the number 230. Draw your blocks below.

3. Use base-10 blocks to show the number 203. Draw your blocks below.

Math Boxes

DATE

1 Write the number that is 100 more.

104 ______

204 ______

304 ______

404 ______

2 Fill in the bubble next to the number word for 38.

- ◯ thirty-eight
- ◯ twenty-eight
- ◯ eighty-three
- ◯ thirty

3 During what part of the day do you leave to go to school?

Circle the correct answer.

A.M. or P.M.

MRB 108-109

4 Draw hands to show 9:15.

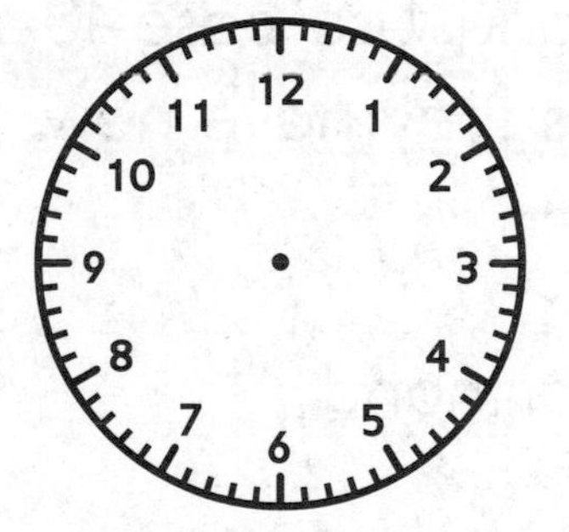

MRB 106-107

5 Place the number 34 in the correct spot on the number line.

30 40

Adding with Base-10 Blocks

Lesson 4-7

DATE

Math Message

Don showed 34 with 3 longs and 4 cubes: 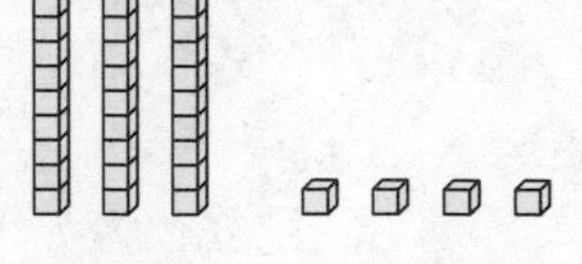

Luke showed 25 with 2 longs and 5 cubes: 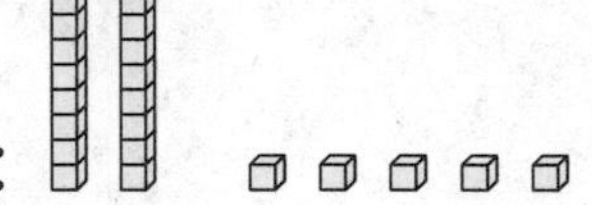

How many longs did Don and Luke have in all together? ________

How many cubes did Don and Luke have in all together? ________

What number is shown by the blocks all together? ______

Luke found one more base-10 cube and traded 10 cubes for 1 long. Use base-10 shorthand to show the blocks they have now.

long = | cube = ■

Use base-10 shorthand.

 Show 15.

Show 35.

What is the total value of the blocks? ______

Show the total value using the fewest blocks:

 Show 27.

Show 25.

What is the total value of the blocks? ______

Show the total value using the fewest blocks:

Practicing with Place Value

Lesson 4-7

DATE

Write the number for each group of base-10 blocks.

1. 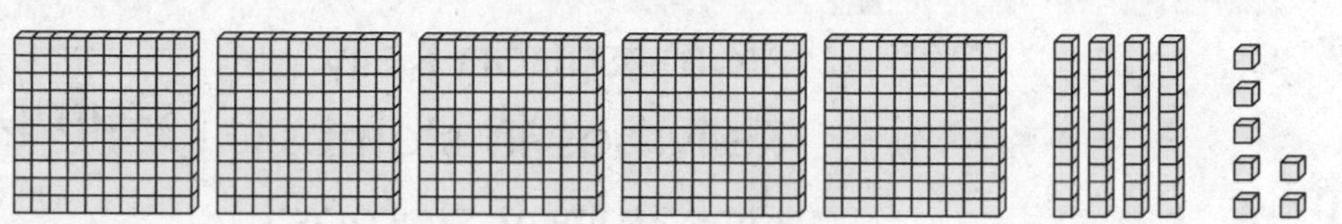 ______

2. 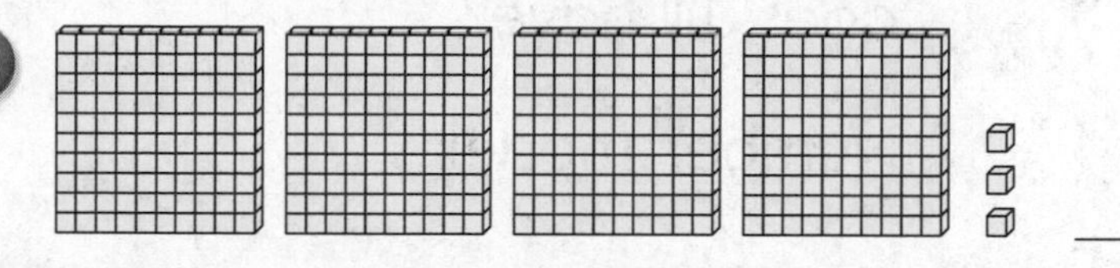 ______

3. Write a number with

   5 in the ones place,

   3 in the hundreds place,

   and 2 in the tens place.

   ______

4. 506

   How many hundreds? ______

   How many tens? ______

   How many ones? ______

Write <, >, or =.

5. 328 ______ 322

6. 122 ______ 102

7. 623 ______ 633

8. Marta wrote 24 to describe the number shown by these base-10 blocks:

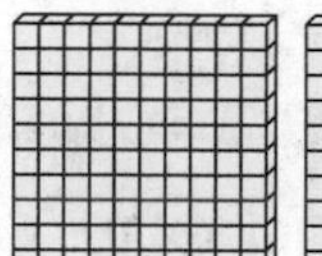
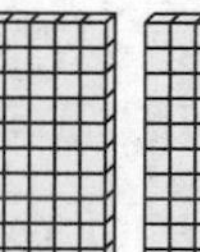

   Do you agree with Marta? ______ Explain your answer.

   ____

   ____

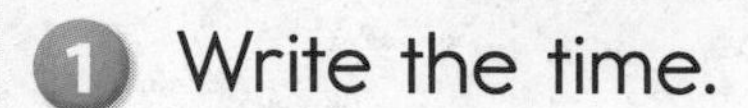

DATE

Math Boxes

1. Write the time.

______ : ______

MRB 106

2. Jill has 9 fewer stamps than Kelly. Kelly has 16 stamps. How many stamps does Jill have?

Unit
stamps

Number model:

____

Answer: ______ stamps

MRB 30–31

3. Use a number grid. What is the distance from . . .

50 to 100? ______

35 to 70? ______

4. Fill in the blanks.

185, 190, ______, 200, ______

5. **Writing/Reasoning** Write a number sentence to show how you found one of the distances in Problem 3.

____

The Foot-Long Foot

Lesson 4-8

DATE

Work with a partner. Follow these steps:

1. Choose an object to measure with your foot-long foot.
2. Each partner writes the name of the object.
3. Each partner measures the length of the object.
4. Talk about your results.
5. When you agree on the measure, circle the sentence that best fits your measure. Record the measure on the line(s).
6. Repeat with other objects.

Name of Object	Length of Object
	It is about ______ feet long. It is between ______ feet and ______ feet long.
	It is about ______ feet long. It is between ______ feet and ______ feet long.
	It is about ______ feet long. It is between ______ feet and ______ feet long.
	It is about ______ feet long. It is between ______ feet and ______ feet long.

Why were your measures and your partner's measures the same for each object? ____________________________________

____

Renaming Numbers

Lesson 4-8

DATE

Work with a partner. Use base-10 shorthand to show your answers. Then write a number model.

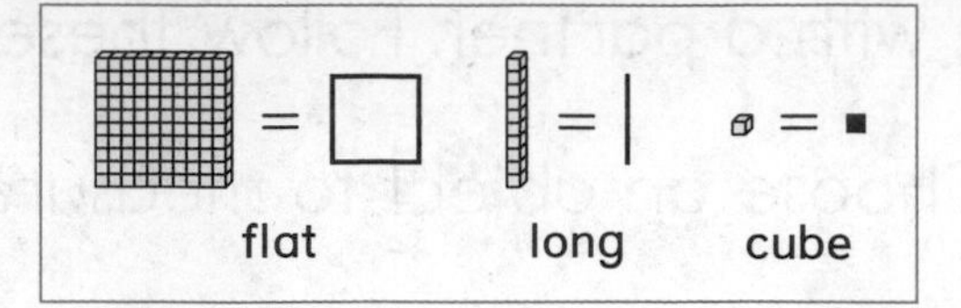

Example: Show 38 with 3 longs and some cubes.

||| ▪▪▪▪▪▪▪▪

30 + 8 = 38

1 Show 38 with 2 longs and some cubes. ______	**2** Show 38 with 1 long and some cubes. ______
3 Show 53 with some longs and 3 cubes. ______	**4** Show 53 with 3 longs and some cubes. ______

Make up your own.

5 Show ______ with ______ longs and some cubes.

6 Show ______ with ______ longs and some cubes.

Math Boxes
Preview for Unit 5

1 Solve.

48 + 10 = ______

32 + ______ = 42

______ + 10 = 16

______ + 75 = 85

2 I bought a banana and a juice box. Each cost 30¢. How much did I spend?

Total	
?	
Part	Part
30¢	30¢

Answer: ______¢

3 Draw the coins you would need to buy a carton of milk for 70¢.

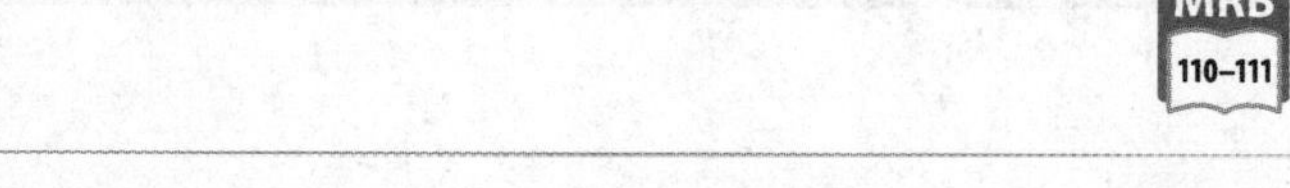

MRB 110–111

4 Solve.

______ = 100 + 1

______ + 100 = 120

201 = ______ + 101

100 + ______ = 171

5 There are 20 airplanes at the airport. Then 8 take off. How many are still on the ground?

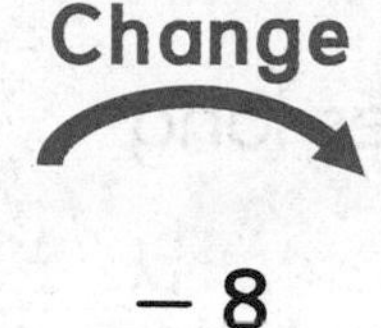

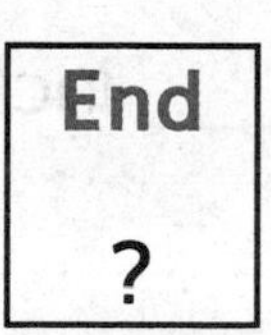

Number model:

____

Answer: ______ airplanes

6 A pencil costs 40¢. Jack paid with two quarters. How much change does he get back? Fill in the bubble next to the correct answer.

- ◯ 25¢
- ◯ 15¢
- ◯ 10¢
- ◯ 30¢

Measuring Lengths

Lesson 4-9

DATE

Use square pattern blocks to measure the length of each object. Record the measures.

Then use a 12-inch ruler to measure the length of each object. Record the measures.

1. About ______ pattern blocks long About ______ inches long

2. About ______ pattern blocks long About ______ inches long

3. About ______ pattern blocks long About ______ inches long

Try This

4. Line up the 3-inch mark on your ruler with the beginning of the paper clip in Problem 1.

   What number is at the end of the paper clip? ______

   How long is the paper clip? About ______ inches long

   How do you know?

   ____

   ____

Measuring a Crooked Path

Lesson 4-9

DATE

1. Use 1-inch long square pattern blocks to measure each part of the ant's path to the nearest inch. Record the measurements in the table.

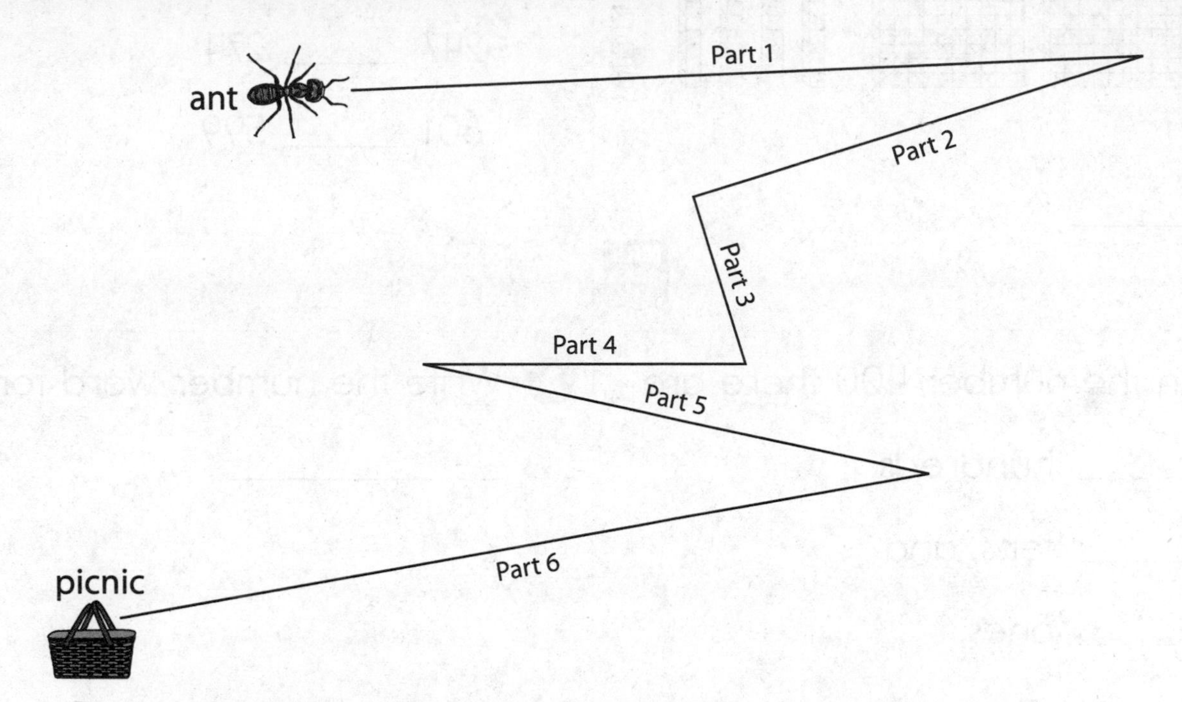

Part	Measurement
Part 1	About ______ inch(es)
Part 2	About ______ inch(es)
Part 3	About ______ inch(es)
Part 4	About ______ inch(es)
Part 5	About ______ inch(es)
Part 6	About ______ inch(es)

2. What is the total length of the path? About ______ inches

Math Boxes

1 Write the number.

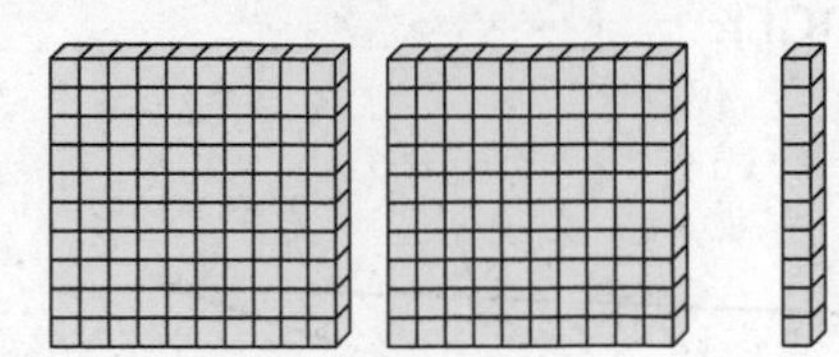

____

2 Write <, >, or =.

415 _______ 315

247 _______ 274

601 _______ 599

3 In the number 400 there are

_______ hundreds,

_______ tens, and

_______ ones.

4 Write the number word for 20.

____

5 **Writing/Reasoning** Draw another way to show the number in Problem 1 with base-10 blocks.

MRB
72

Measuring Lengths

Lesson 4-10

DATE

Measure each object in inches and in centimeters.
Record your measures.

1 

About ______ inches long About ______ centimeters long

 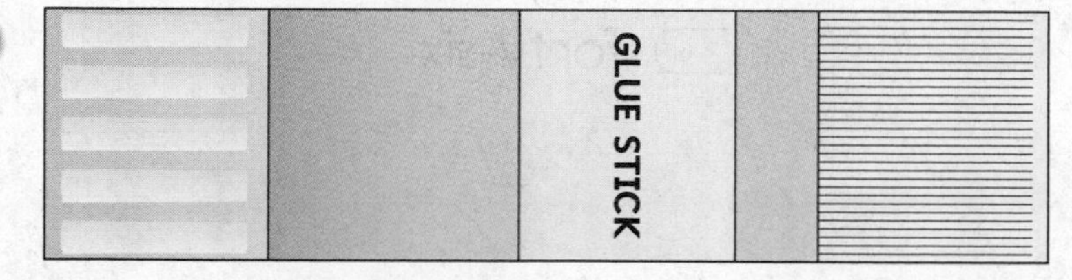

About ______ inches long About ______ centimeters long

Pick short objects to measure. Write the name of each object or draw a picture of it. Then measure the objects in inches and in centimeters. Record your measures.

3

About ______ inches About ______ centimeters

4

About ______ inches About ______ centimeters

Math Boxes

Lesson 4-10

DATE

1. Write the number that is 100 more.

123 ______

607 ______

399 ______

801 ______

2. Fill in the bubble next to the number word for 46.

◯ forty-four
◯ forty
◯ sixty-four
◯ forty-six

3. During what part of the day do you eat supper?

Circle the correct answer.

A.M. or P.M.

MRB 108–109

4. What time does the clock show?

______ : ______

5. Place the number 72 in the correct spot on the number line.

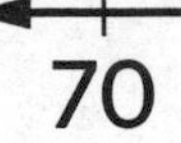

70 80

Matching Subtraction Facts to Strategies

Lesson 4-11

DATE

Subtraction Strategies				
Think Addition	Count Up	Count Back	Go Through 10	Other

Solving Number Stories

Lesson 4-11

DATE

Solve. Use your number grid to help.

1. Joey had 40 baseball cards. He received 10 more. How many does he have now?

   _______ baseball cards

2. Nathan had a box of books. He borrowed 10 more from his brother. Now he has 68 books. How many did he start with?

   _______ books

3. Lisa had 34 beads. Her friends gave her more beads. Now she has 44 beads. How many beads did her friends give her?

   _______ beads

4. Ameer had a bag of crayons. He got a set of 10 more at the store. Now he has 82. How many crayons did he have before?

   _______ crayons

5. Briana had 126 stickers in her collection. She bought 10 more. How many stickers does she have now?

   _______ stickers

Math Boxes

Lesson 4-11

DATE

1 Write a number with 5 in the ones place, 2 in the tens place, and 7 in the hundreds place.

____

2 Write <, >, or =.

319 ______ 219

645 ______ 654

789 ______ 901

MRB 74–75

3 In the number 800 there are

______ hundreds.

______ tens.

______ ones.

4 Write the number word for 8.

____

Write the number word for 70.

____

5 **Writing/Reasoning** For Problem 1, suppose your teacher asked your class to write the number in expanded form. Your friend Cassie wrote 700 + 2 + 5. Do you agree? Why or why not?

____

____

____

MRB 72

Math Boxes
Preview for Unit 5

Lesson 4-12

DATE

Math Boxes

1. Solve.

   ______ + 78 = 88

   62 + 10 = ______

   54 + ______ = 64

   ______ + 10 = 19

2. I bought an orange for 42¢ and an apple for 20¢.

   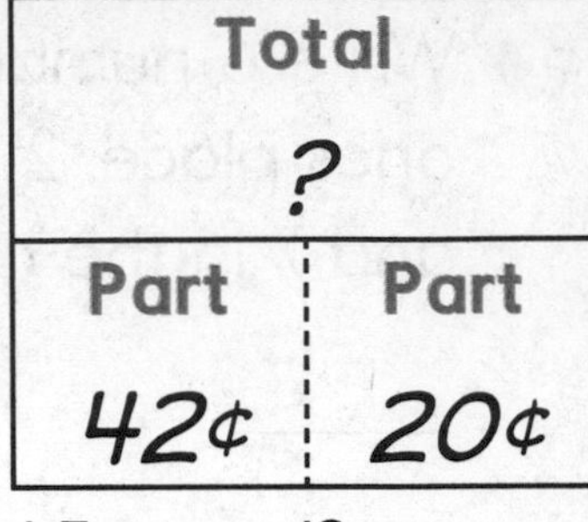

   How much did I spend?

   Answer: ______

3. Draw the coins you would need to buy a toy car for 85¢.

   MRB 110-111

4. Add.

   100 + 100 = ______

   100 + ______ = 159

   300 = ______ + 200

   ______ + 100 = 165

5. Natalia sold 25 books at her garage sale on Saturday. She sold 15 more on Sunday. How many did she sell in all?

   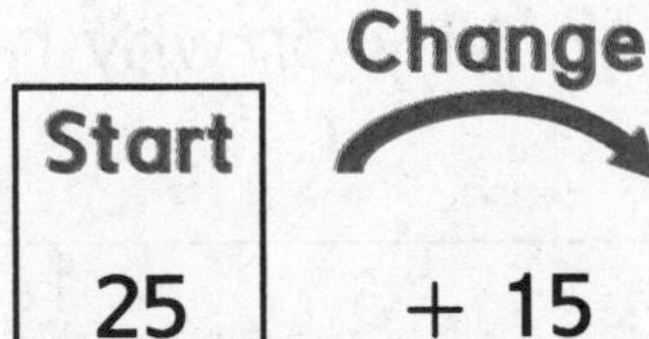

   Number model:

   ____

   Answer: ______ books

   MRB 28

6. An eraser costs 70¢. Jack pays with three quarters. How much change does he get? Fill in the bubble next to the correct amount of change.

   ◯ 25¢
   ◯ 15¢
   ◯ 5¢
   ◯ 10¢

Addition Facts Inventory Record, Part 1

Lesson 3-3

DATE

Addition Fact	Know It	Don't Know It	How I Can Figure It Out
10 + 3			
4 + 6			
7 + 7			
3 + 2			
3 + 4			
10 + 2			
9 + 9			
2 + 2			
4 + 10			
6 + 6			
10 + 5			
5 + 2			
3 + 7			
4 + 4			

Addition Facts Inventory Record, Part 1 (continued)

DATE

Addition Fact	Know It	Don't Know It	How I Can Figure It Out
10 + 6			
5 + 5			
8 + 10			
2 + 4			
3 + 3			
10 + 7			
7 + 2			
8 + 8			
9 + 10			
9 + 2			
2 + 6			
10 + 10			
2 + 8			

Addition Facts Inventory Record, Part 2

Lesson 3-8

DATE

Addition Fact	Know It	Don't Know It	How I Can Figure It Out
3 + 5			
3 + 6			
3 + 8			
3 + 9			
4 + 5			
4 + 7			
4 + 8			
4 + 9			
5 + 6			

Addition Facts Inventory Record, Part 2 (continued)

Lesson 3-8

DATE

Addition Fact	Know It	Don't Know It	How I Can Figure It Out
$5 + 7$			
$5 + 8$			
$5 + 9$			
$6 + 7$			
$6 + 8$			
$6 + 9$			
$7 + 8$			
$7 + 9$			
$8 + 9$			

Notes

Fact Triangles: Set 1

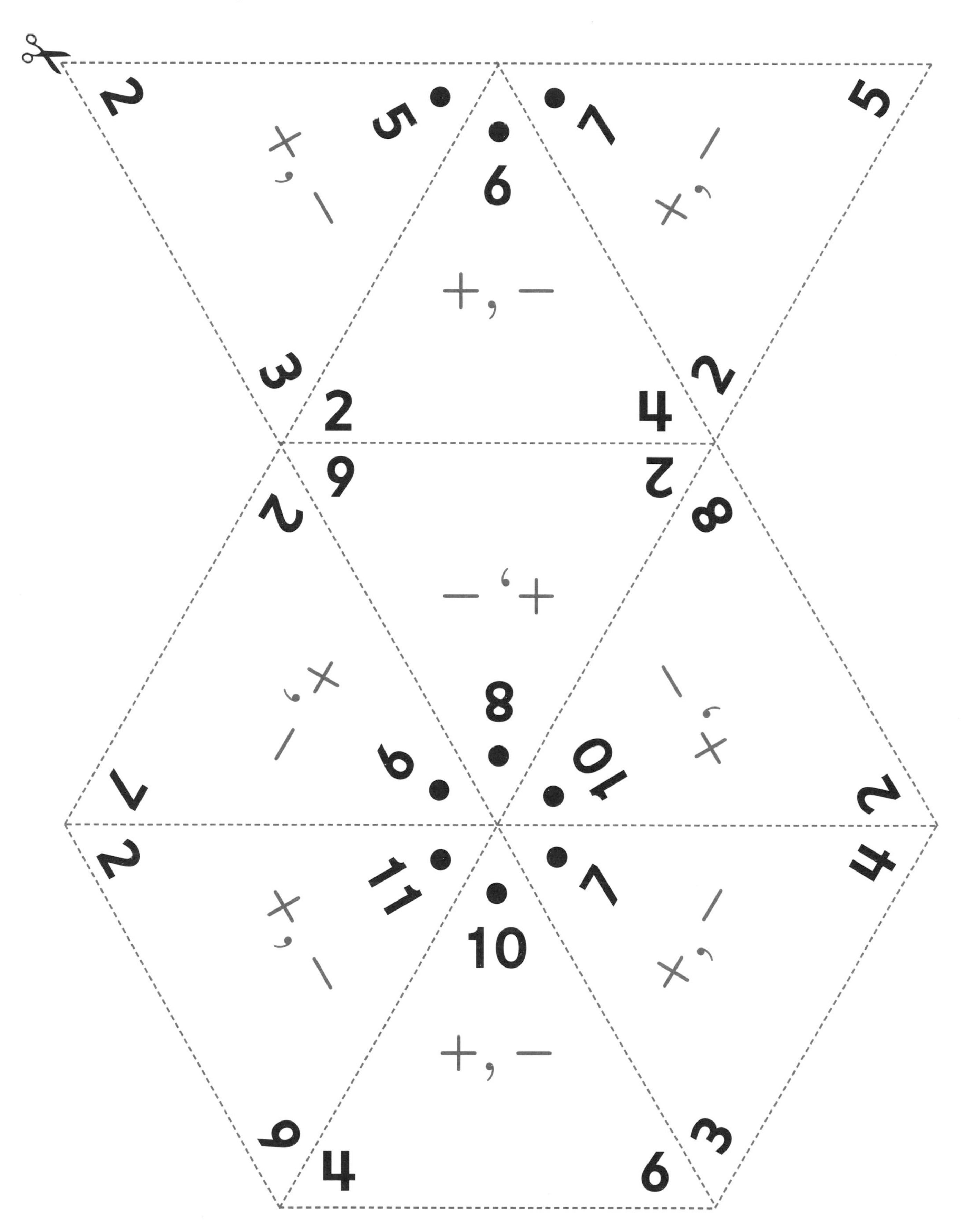

Fact Triangles: Set 1 (continued)

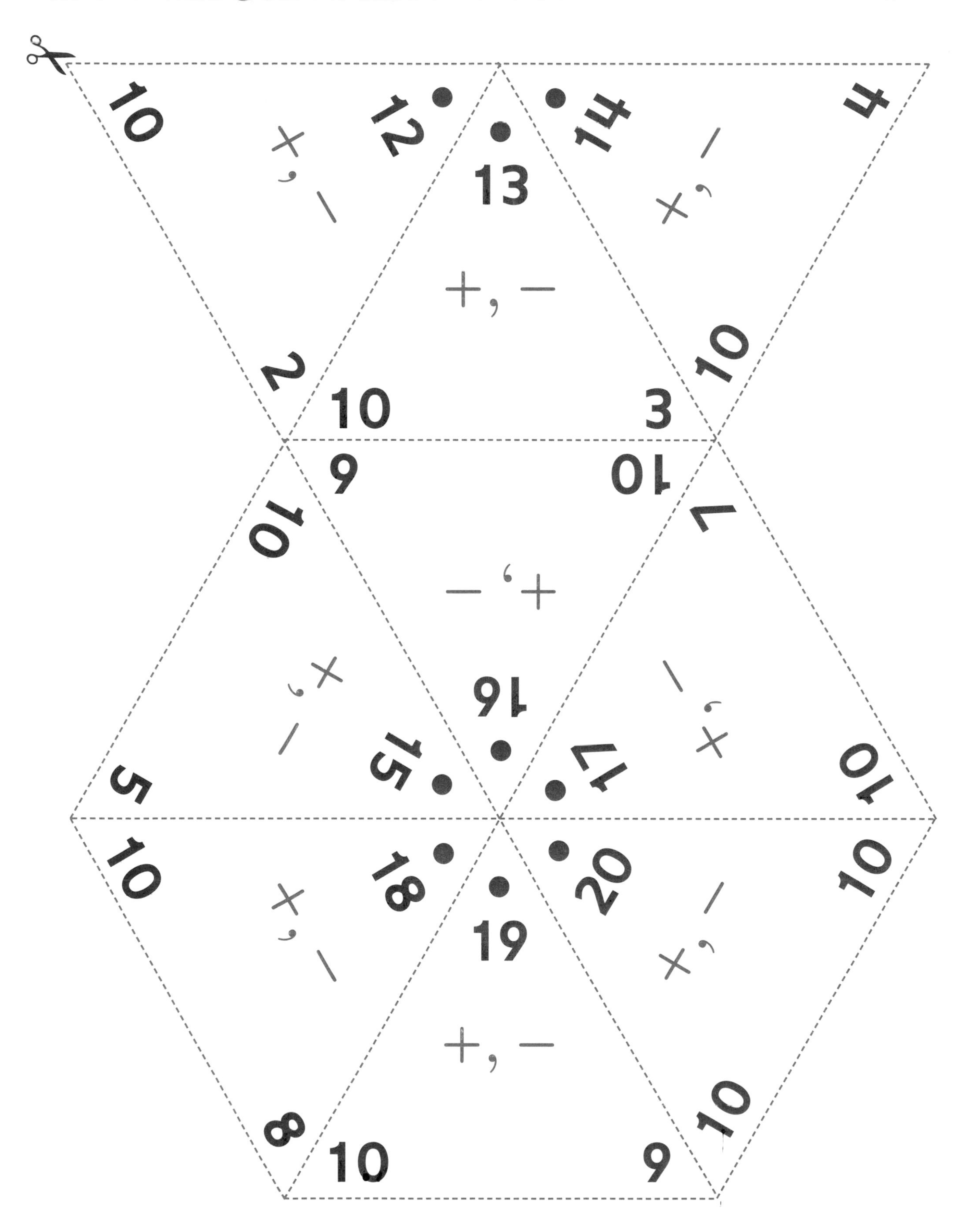

Fact Triangles: Set 1 (continued)

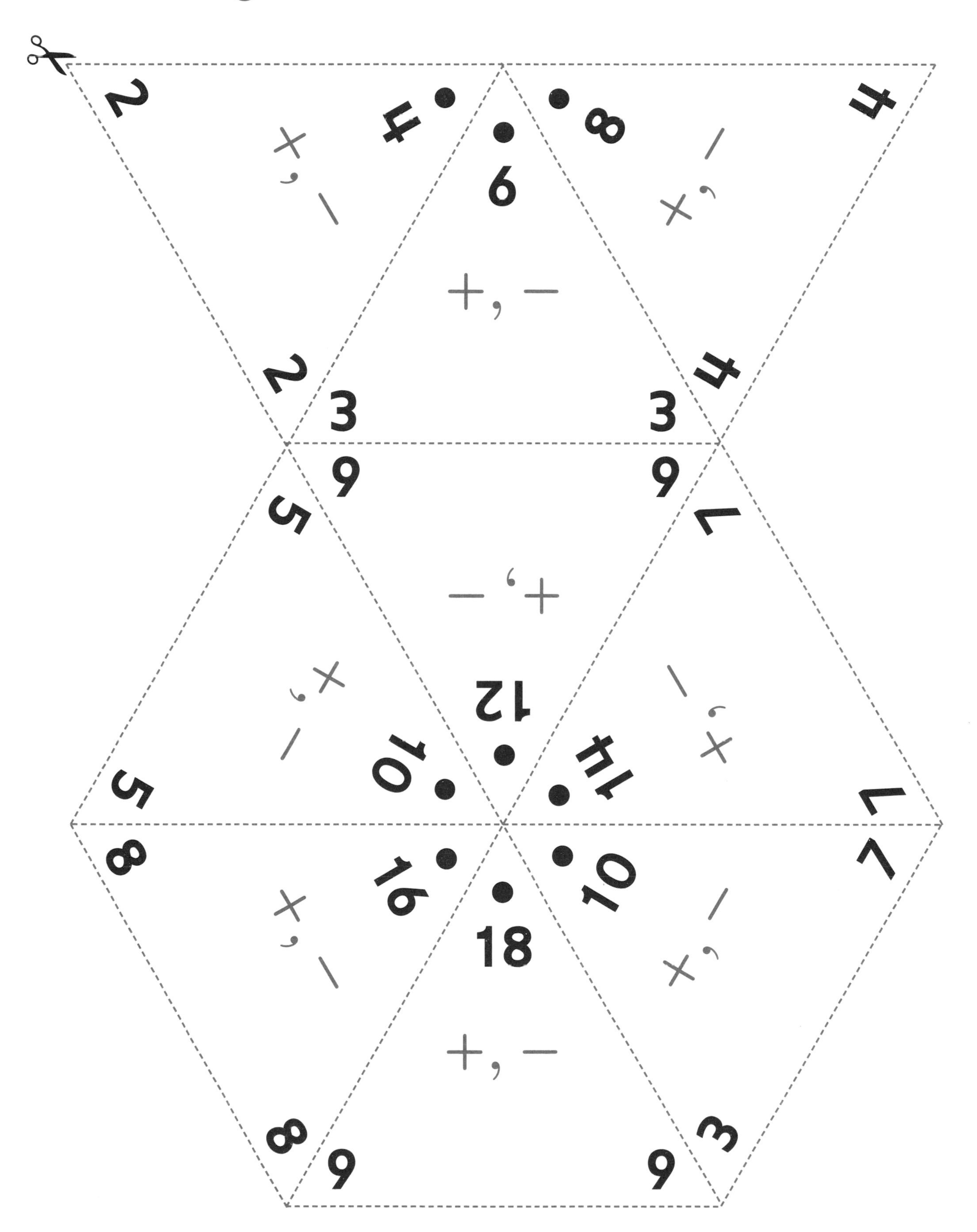

Fact Triangles: Set 2

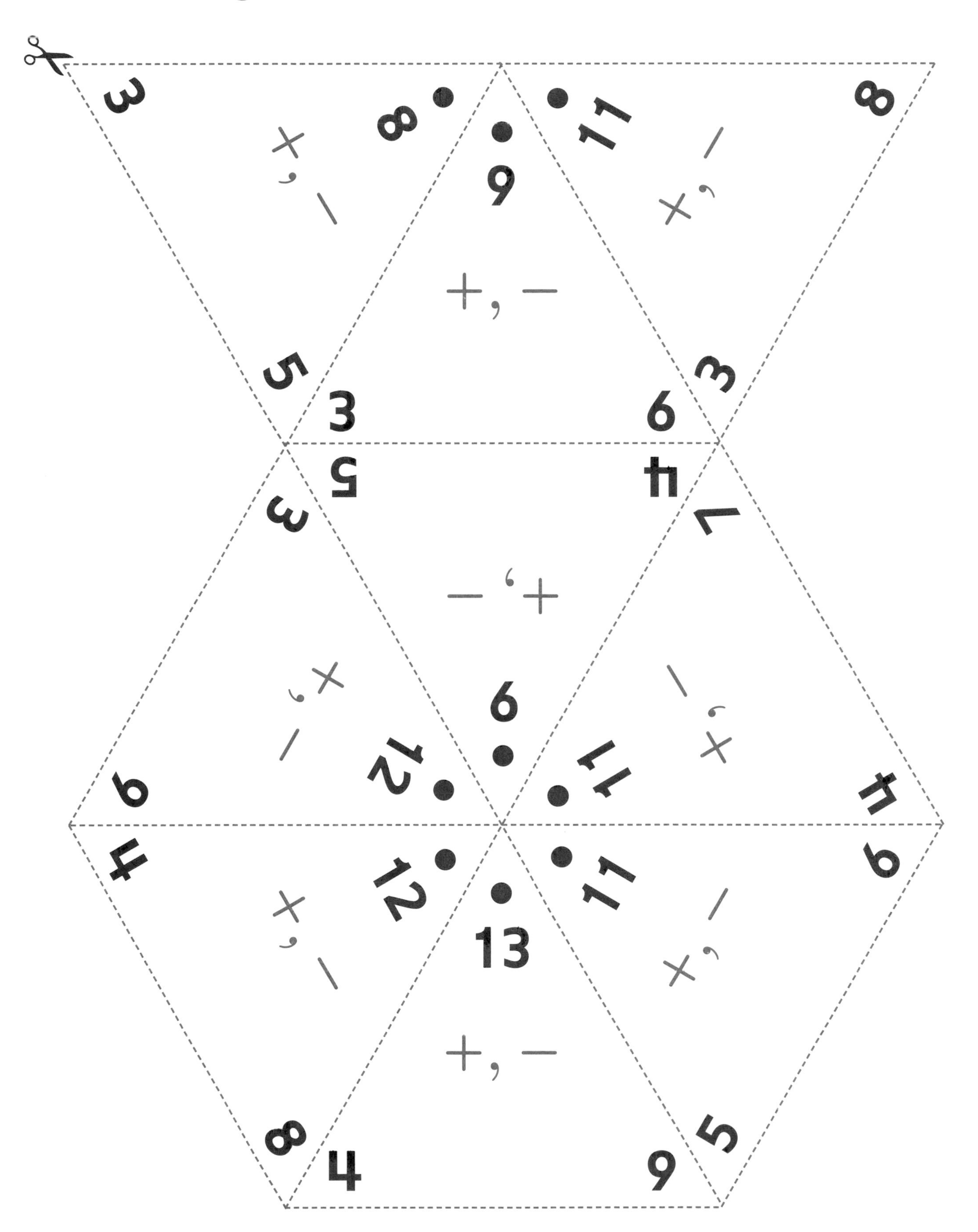

Fact Triangles: Set 2 (continued)

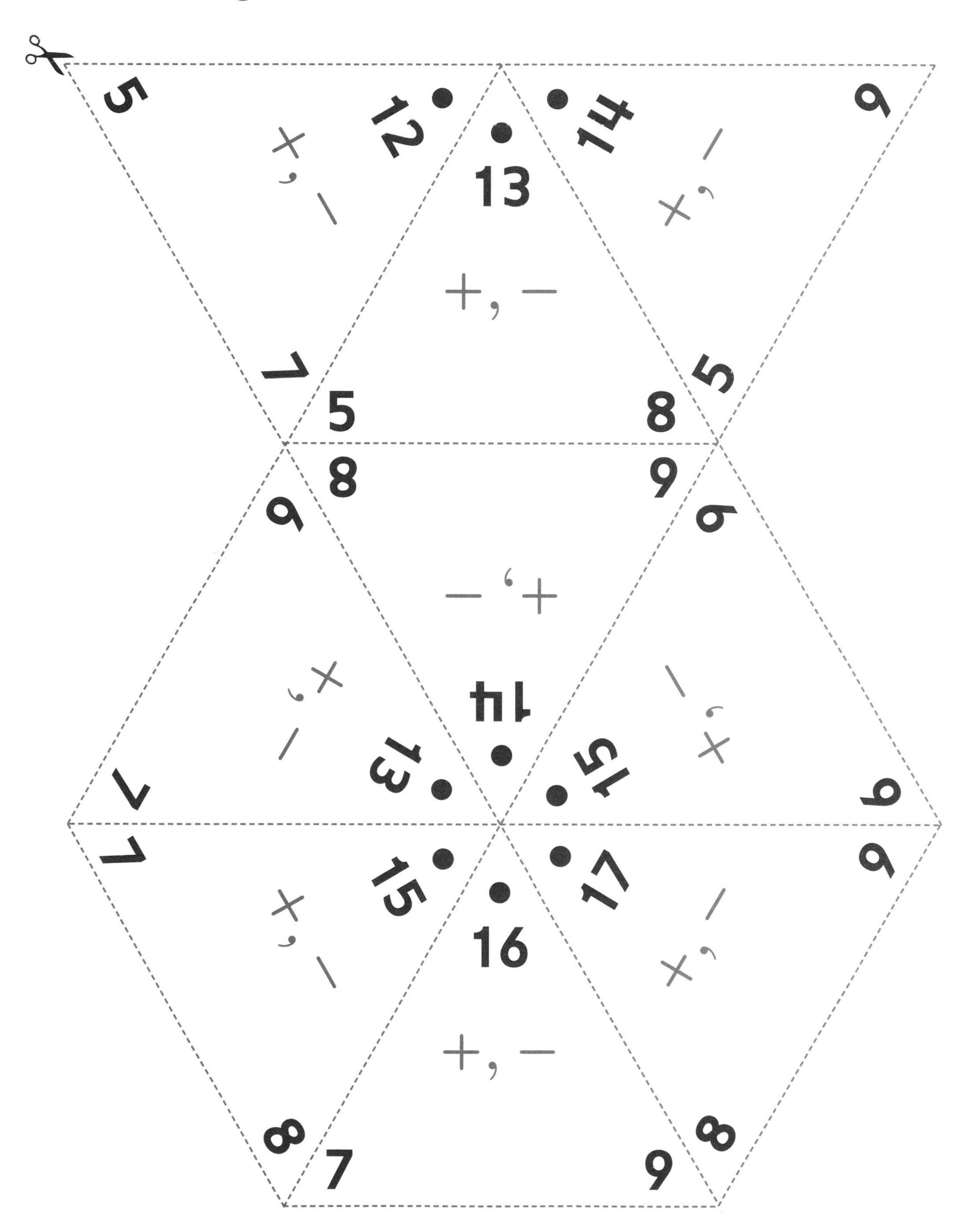

A 12-Inch (Foot) Ruler and 10-Centimeter Ruler

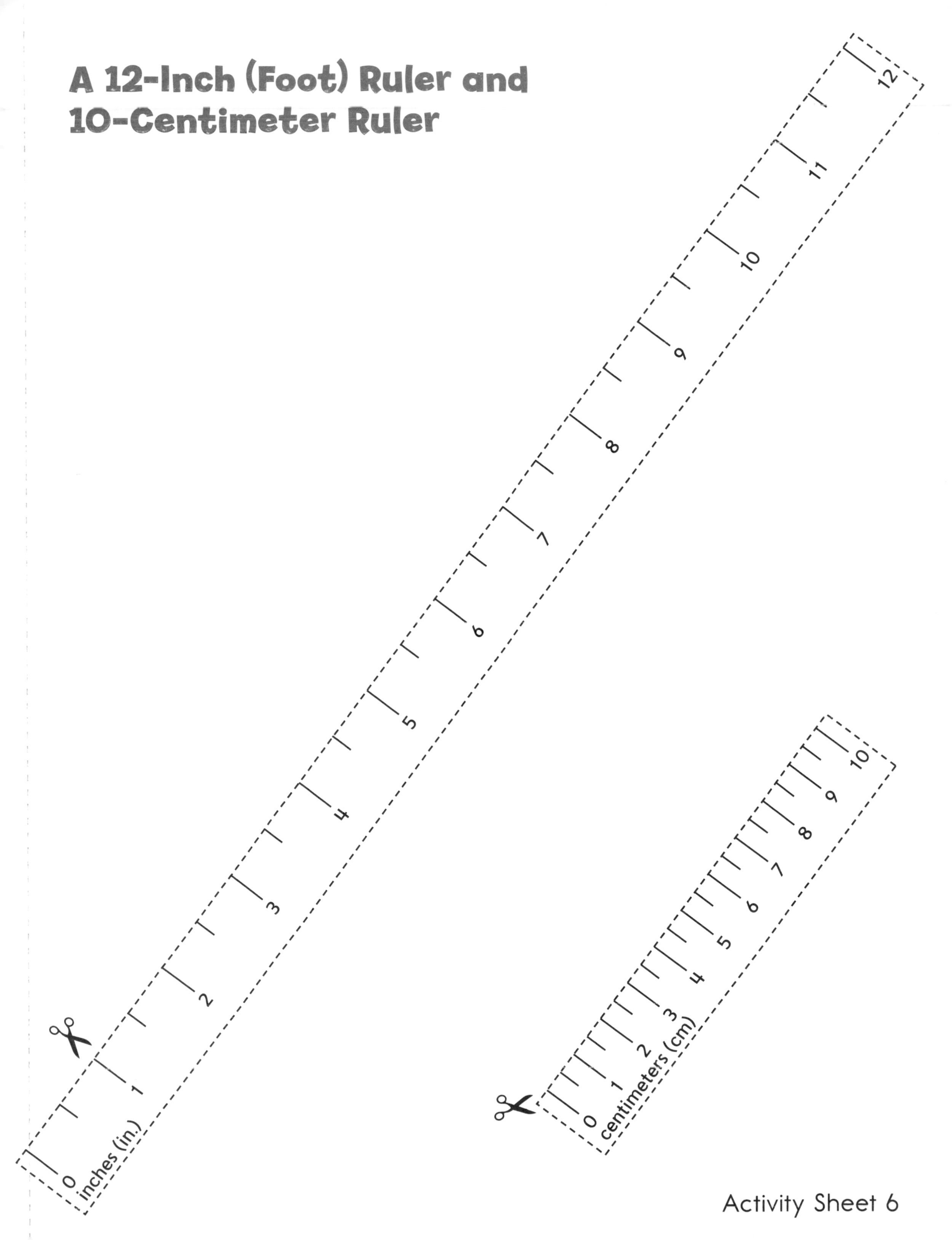